职业教育烹饪（餐饮）类专业"以工作过程为导向"
课程改革"纸数一体化"系列精品教材

XICAN PENGREN YINGYU

西餐烹饪英语

主　编　张艳红
副主编　赵　静　刘欣雨
参　编（以编写章节顺序排列）
　　　　　钱　俊　贾尚谕　李　蕊　王　茜
　　　　　孙文红　王　静　侯雪璘

华中科技大学出版社
http://www.hustp.com
中国·武汉

内 容 简 介

本教材为职业教育烹饪(餐饮)类专业"以工作过程为导向"课程改革"纸数一体化"系列精品教材。

本教材共10个单元,内容包括厨房简介、厨房设备、工具和餐具、调味品和香料、蔬菜、水果和坚果、肉类、海产品、调味汁和汤、油酥糕点和西饼。本教材用生动形象、通俗易懂的图片、对话、故事介绍西餐烹饪英语相关知识,使教材更具有可读性,同时配备英文听力材料、教学课件等丰富的数字教学资源。

本教材适合职业教育西餐烹饪等相关专业使用,同时也可作为餐饮企业员工的培训教材。

图书在版编目(CIP)数据

西餐烹饪英语/张艳红主编. —武汉:华中科技大学出版社,2020.8
ISBN 978-7-5680-6428-6

Ⅰ.①西… Ⅱ.①张… Ⅲ.①西式菜肴-烹饪-英语-中等专业学校-教材 Ⅳ.①TS972.118

中国版本图书馆 CIP 数据核字(2020)第 154838 号

西餐烹饪英语
Xican Pengren Yingyu

张艳红　主编

策划编辑:汪飒婷	
责任编辑:汪飒婷　马梦雪	
封面设计:原色设计	
责任校对:阮　敏	
责任监印:周治超	
出版发行:华中科技大学出版社(中国·武汉)	电话:(027)81321913
武汉市东湖新技术开发区华工科技园	邮编:430223
录　　排:华中科技大学惠友文印中心	
印　　刷:武汉科源印刷设计有限公司	
开　　本:889mm×1194mm　1/16	
印　　张:10.5	
字　　数:243 千字	
版　　次:2020 年 8 月第 1 版第 1 次印刷	
定　　价:49.80 元	

本书若有印装质量问题,请向出版社营销中心调换
全国免费服务热线:400-6679-118　竭诚为您服务
版权所有　侵权必究

职业教育烹饪（餐饮）类专业"以工作过程为导向"
课程改革"纸数一体化"系列精品教材

编委会

主任委员

郭延峰　北京市劲松职业高中校长
董振祥　大董餐饮投资有限公司董事长

副主任委员

刘雪峰　山东省城市服务技师学院中餐学院院长
刘铁锁　北京市延庆区第一职业学校校长
刘慧金　北京新城职业学校校长
赵　军　唐山市第一职业中专校长
李雪梅　张家口市职业技术教育中心校长
杨兴福　禄劝彝族苗族自治县职业高级中学校长
刘新云　大董餐饮投资有限公司人力资源总监

委　员

王为民　张晶京　范春玥　杨　辉　魏春龙
赵　静　向　军　刘寿华　吴玉忠　王蛰明
陈　清　侯广旭　罗睿欣　单　蕊

总序
FOREWORD

职业教育作为一种类型教育，其本质特征，诚如我国职业教育界学者姜大源教授提出的"跨界论"：职业教育是一种跨越职场和学场的"跨界"教育。

习近平总书记在十九大报告中指出，要"完善职业教育和培训体系，深化产教融合、校企合作"，为职业教育的改革发展提出了明确要求。按照职业教育"五个对接"的要求，即专业设置与产业需求对接、专业课程内容与职业标准对接、教学过程与生产过程对接、学历证书与职业资格证书对接、职业教育与终身学习对接，深化人才培养模式改革，完善专业课程体系，是职业教育发展的应然之路。

国务院印发的《国家职业教育改革实施方案》(国发〔2019〕4号)中强调，要借鉴"双元制"等模式，校企共同研究制定人才培养方案，及时将新技术、新工艺、新规范纳入教学标准和教学内容，建设一大批校企"双元"合作开发的国家规划教材，倡导使用新型活页式、工作手册式教材并配套开发信息化资源。

北京市劲松职业高中贯彻落实国家职业教育改革发展的方针和要求，与大董餐饮投资有限公司及20余家星级酒店深度合作，并联合北京、山东、河北等一批兄弟院校，历时两年，共同编写完成了这套"职业教育烹饪(餐饮)类专业'以工作过程为导向'课程改革'纸数一体化'系列精品教材"。教材编写经历了行业企业调研、人才培养方案修订、课程体系重构、课程标准修订、课程内容丰富与完善、数字资源开发与建设几个过程。其间，以北京市劲松职业高中为首的编写团队在十余年"以工作过程为导向"的课程改革基础上，根据行业新技术、新工艺、新标准以及职业教育新形势、新要求、新特点，以"跨界""整合"为学理支撑，产教深度融合，校企密切合作，审纲、审稿、论证、修改、完善，最终形成了本套教材。在编写过程中，编委会一直坚持科研引领，2018年12月，"中餐烹饪专业'三级融合'综合实训项目体系开发与实践"获得国家级教学成果奖二等奖，以培养综合职业能力为目标的"综合实训"项目在中餐烹饪、西餐烹饪、高星级酒店运营与管理专业的专业核心课程中均有体现。凸显"跨界""整合"特征的《烹饪语文》《烹饪数学》《中餐烹饪英语》《烹饪体育》等系列公共基础课职业模块教材是本套教材的另一特色和亮点。大董餐饮投资有限公司主持编写的相关教材,更是让本套教材锦上添花。

本套教材在课程开发基础上,立足于烹饪(餐饮)类复合型、创新型人才培养,以就业为导向,以学生为主体,注重"做中学""做中教",主要体现了以下特色。

1. 依据现代烹饪行业岗位能力要求,开发课程体系

遵循"以工作过程为导向"的课程改革理念,按照现代烹饪岗位能力要求,确定典型工作任务,并在此基础上对实际工作任务和内容进行教学化处理、加工与转化,开发出基于工作过程的理实一体化课程体系,让学生在真实的工作环境中,习得知识,掌握技能,培养综合职业能力。

2. 按照工作过程系统化的课程开发方法,设置学习单元

根据工作过程系统化的课程开发方法,以职业能力为主线,以岗位典型工作任务或案例为载体,按照由易到难、由基础到综合的逻辑顺序设置三个以上学习单元,体现了学习内容序化的系统性。

3. 对接现代烹饪行业和企业的职业标准,确定评价标准

针对现代烹饪行业的人才需求,融入现代烹饪企业岗位工作要求,对接行业和企业标准,培养学生的实际工作能力。在理实一体教学层面,夯实学生技能基础。在学习成果评价方面,融合烹饪职业技能鉴定标准,强化综合职业能力培养与评价。

4. 适应"互联网+"时代特点,开发活页式"纸数一体化"教材

专业核心课程的教材按新型活页式、工作手册式设计,图文并茂,并配套开发了整套数字资源,如关键技能操作视频、微课、课件、试题及相关拓展知识等,学生扫二维码即可自主学习。活页式及"纸数一体化"设计符合新时期学生学习特点。

本套教材不仅适合于职业院校餐饮类学生教学使用,还适用于相关社会职业技能培训。数字资源既可用于学生自学,还可用于教师教学。

本套教材是深度产教融合、校企合作的产物,是十余年"以工作过程为导向"的课程改革成果,是新时期职教复合型、创新型人才培养的重要载体。教材凝聚了众多行业企业专家、一线高技能人才、具有丰富教学经验的教师及各学校领导的心血。教材的出版必将极大地丰富北京市劲松职业高中餐饮服务特色高水平骨干专业群及大董餐饮文化学院建设内涵,提升专业群建设品质,也必将为其他兄弟院校的专业建设及人才培养提供重要支撑,同时,本套教材也是落实国家"三教改革"要求的积极探索,教材中的不足之处还请各位专家、同仁批评指正!我们也将在使用中不断总结、改进,期待本套教材拥有良好的育人效果。

职业教育烹饪(餐饮)类专业"以工作过程为导向"课程改革
"纸数一体化"系列精品教材编委会

前言

《西餐烹饪英语》是中等职业学校烹饪专业的英语教材,也可以作为西餐从业人员和西餐美食爱好者的自学教材。本教材遵循以就业为导向,以职业能力为本位,将行业知识和职业技能渗透在英语教学中,旨在培养学生掌握必要的专业词汇,熟悉餐饮从业人员的基本技能,并通过课后练习等环节使学生可以检测自己的学习效果。本教材分成10个单元,适用于72个学时的课程教学安排。

本教材在编写时突出以下几个特色。

1. 实践性

以一位高三毕业生Jack到饭店实习中的所见所听所感为主线,以"情景教学为基石,任务为驱动,活动为内容"的方式编写,突出实践技能,强调应用,内容由浅及深,层层递进,不断提高学生的参与程度和学习兴趣,保证学习目标的实现。

2. 专业性

在参阅大量国内外西餐英语教材和深入企业调研的基础上编写而成,每个章节都由行业专家和一线西餐教师参与,在专业知识方面给予直接的指导。

3. 趣味性

贯穿"以学生为中心"的教学理念,设计了丰富多彩的课堂教学活动,使学生在真实的情景下进行听说读写的训练;采用大量的图片激发学生的学生兴趣,从视觉和听觉上不断吸引学生关注,激发学生学习的潜能。

本教材由北京市朝阳区教育研究中心张艳红担任主编,北京市劲松职业高中赵静、刘欣雨担任副主编,北京市劲松职业高中钱俊、贾尚谕、李蕊、王茜、孙文红、王静和侯雪璘参编。具体编写分工:Unit 1由钱俊负责编写;Unit 2由贾尚谕负责编写;Unit 3由刘欣雨负责编写;Unit 4由李蕊负责编写;Unit 5由王茜负责编写;Unit 6由孙文红负责编写;Unit 7由王静负责编写;Unit 8由张艳红负责编写;Unit 9由赵静负责编写;Unit 10由侯雪璘负责编写。张艳红老师负责全书的统稿工作。

本教材在编写过程中,广泛吸取了国内外现有的研究成果,参考了张艳红老师编写的西餐烹饪英语教材,引用了其中有关文献材料。本教材同时得到了烹饪专家和华中科技大学出版社编辑的指导和帮助,一并表示衷心感谢!由于学识有限,书中不妥之处敬请读者批评指正。

　　为了方便教学,本教材还配有链接教学课件、习题答案和英语听力材料等相关教学资源的二维码。

1	**Unit 1**	**Kitchen Introduction**	
3	Task 1	Words & expressions 词汇	
5	Task 2	Chef's uniform, position, hygiene & safety 厨师的工作服、职责、卫生和厨房安全	
8	Task 3	Other hygiene and safety rules in the kitchen 其他卫生安全常识	
17	**Unit 2**	**Kitchen Equipment**	
19	Task 1	Words & expressions 词汇	
21	Task 2	Cooking equipment 烹饪设备	
24	Task 3	Auxiliary equipment 辅助设备	
31	**Unit 3**	**Tools & Utensils**	
33	Task 1	Words & expressions 词汇	
35	Task 2	Cookware & tableware 烹饪工具和餐具	
38	Task 3	Cutlery 刀具	
45	**Unit 4**	**Condiments & Spices**	
47	Task 1	Words & expressions 词汇	
49	Task 2	Primary processing 初加工	
52	Task 3	Making dishes 制作菜肴	
59	**Unit 5**	**Vegetables**	
61	Task 1	Words & expressions 词汇	
63	Task 2	Primary processing 初加工	

Task 3	Making typical dishes 特色菜肴制作	66

Unit 6 Fruits & Nuts   75

Task 1	Words & expressions 词汇	77
Task 2	Primary processing 初加工	79
Task 3	Making typical dishes 特色菜肴制作	83

Unit 7 Meat   91

Task 1	Words & expressions 词汇	93
Task 2	Primary processing 初加工	96
Task 3	Making typical dishes 特色菜肴制作	99

Unit 8 Seafood   107

Task 1	Words & expressions 词汇	109
Task 2	Primary processing 初加工	112
Task 3	Making typical dishes 特色菜肴制作	115

Unit 9 Sauce & Soup   125

Task 1	Words & expressions 词汇	127
Task 2	Sauce making 少司的制作	129
Task 3	Making soup 汤菜制作	132

Unit 10 Pastry & Bakery   141

Task 1	Words & expressions 词汇	143
Task 2	Prebaking 预烘	146
Task 3	Baking 烘焙	149

Unit 1
Kitchen Introduction

You will be able to:

1. know the chef's positions in the kitchen;
2. describe job responsibilities in the kitchen;
3. strengthen the awareness of occupational health and safety in the kitchen.

Unit introduction 单元介绍

作为西餐专业学生,对西餐厨房的初步了解十分重要,本单元我们将会学到厨师岗位名称、岗位职责、加工间名称和厨房卫生、安全等相关的英文单词及表达方法。

扫码看课件

Thinking map（思维导图）

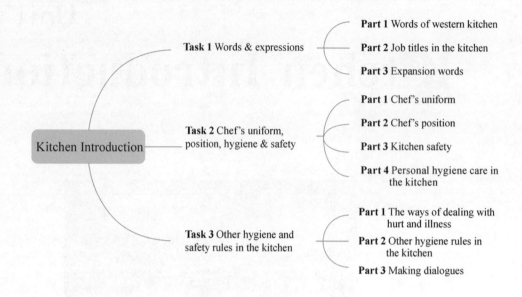

Warming up

Activity 1 Look and match. 看单词，连线。

chef   hotel   cook   trainee   kitchen

外教有声

Activity 2 Listen and tick. 听录音，勾出汤姆的职业。

Task 1 Words & expressions 词汇

Part 1 Words of western kitchen. 西餐各厨房单词。

 Activity 1 Listen and number. 听录音，标出正确顺序。

(　) main kitchen 主厨房　　　　　　(　) banquet kitchen 宴会厨房
(　) hot kitchen 热菜厨房　　　　　　(　) cold kitchen 冷菜厨房
(　) butchery 肉房　　　　　　　　　(　) Italian kitchen 意大利厨房
(　) pastry & bakery 包饼房　　　　　(　) coffee shop kitchen 咖啡厅厨房

Activity 2 Look and translate. 单词互译。

 主厨房_____　　　　 banquet kitchen _____

 冷菜厨房_____　　　 肉房_____

 Italian kitchen _____　 热菜厨房_____

 coffee shop kitchen _____　 pastry & bakery _____

Part 2 Job titles in the kitchen. 厨师岗位单词。

 Activity 1 Read and translate. 读单词并翻译厨房工作岗位。

executive chef 行政总厨　　　　　　　executive sous-chef 行政副总厨
coffee shop kitchen chef 咖啡厅厨师长　cold kitchen chef 冷菜厨房厨师长
banquet kitchen chef 宴会厨房厨师长　　pastry & bakery chef 包饼房厨师长
Italian kitchen chef 意大利厨房厨师长　butchery chef 肉房厨师长

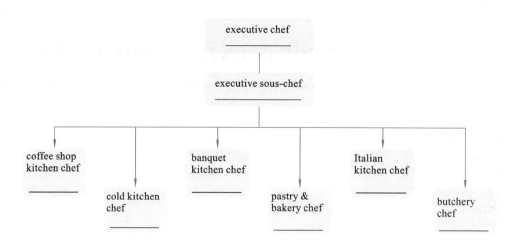

Activity 2 Look and write. 根据描述写出厨师岗位名称。

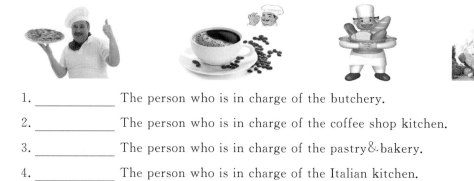

1. _____ The person who is in charge of the butchery.
2. _____ The person who is in charge of the coffee shop kitchen.
3. _____ The person who is in charge of the pastry & bakery.
4. _____ The person who is in charge of the Italian kitchen.

Part 3 Expansion words. 拓展词汇。

Activity 1 Look and write. 根据所给信息写出包饼房各级岗位的英文名称。

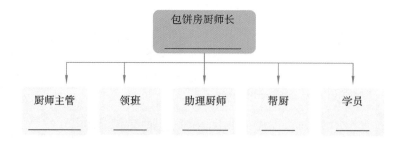

pastry & bakery chef   commis chef   chef de partie   demi chef   kitchen helper   trainee

Activity 2 Listen and write. 听录音写单词。

_____
_____

Unit 1 Kitchen Introduction

Activity 3 Fill in the missing letters and make a self-evaluation. 补全单词并做自我评价。

1. ch __ f
2. banqu __ t
3. k __ tchen
4. c __ mmis ch __ f
5. train ___
6. butch ___ y
7. hot __ l
8. c ___ k
9. p __ stry & bak ___ y
10. Ital __ an

Assessment: If you can write 8-10 words, you are perfect.

If you can write 4-7 words, you are good.

If you can write 1-3 words, you will try again.

Task 2 Chef's uniform, position, hygiene & safety
厨师的工作服、职责、卫生和厨房安全

Part 1 Chef's uniform. 厨师的工作服。

Activity Learn and write these words about uniform. 学习单词，看图写出相应的英文。

chef hat 厨师帽　　　tie 三角巾　　　towel 手巾　　　apron 围裙
jacket 上衣　　　　　shoes 鞋　　　　plaid pants 格子裤

Part 2　Chef's position. 厨师职位。

Activity 1 Look and choose. 看图并选择。

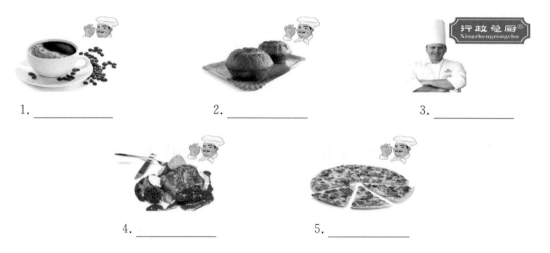

1. _____　2. _____　3. _____

4. _____　5. _____

A. The executive chef is the leader of all sous-chefs.

B. The butchery chef is cooking beef steak in the butchery.

C. The Italian kitchen chef is in the Italian kitchen.

D. The coffee shop kitchen chef is working in the coffee garden.

E. The pastry & bakery chef is in charge of making cakes.

 Activity 2 Listen and fill in the blanks. 听录音，完成对话。

| A. what he does in the kitchen | B. Welcome to our kitchen |
| C. show you around | D. in charge of the butchery |

Chef: Good morning, Jack. _____.

Jack: Good morning, Nice to meet you.

Chef: It's your first day here, let me _____. This way, please. Here is our executive chef office, cold kitchen, and hot kitchen. Oh, this is Mr. Smith. He is a butcher.

Jack: Could you tell me _____?

Chef: He is _____. It's a very important job in the western kitchen.

Jack: I see. Thank you very much.

Chef: You are welcome.

Part 3　Kitchen safety. 厨房安全。

Activity 1 Match the picture with the right phrase. 请连接图片和词组。

knife safety 安全用刀　　fire prevention 防止火灾　　no smoking 禁止吸烟
electric safety 安全用电　　gas prevention 防止漏气

Activity 2 Look and translate. 看图，翻译英文。

1. Do not bring food into the working section.
　_____

2. Do not lend kitchen tools to others.
　_____

3. Keep the knife well away from the table edge.
　_____

4. Clean kitchen facilities and utensils regularly.
　_____

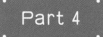

Part 4   Personal hygiene care in the kitchen. 厨房个人卫生。

Activity 1 Look at the pictures of washing hands. 看厨师洗手的过程。

wet   rub   rinse   hold   dry

外教有声

Activity 2 Listen and complete the dialogue. 听录音，参考 **Activity 1** 完成对话。

A. rinse hands with water

B. dry hands with a clean towel

C. rub hands for 20 seconds

D. hold water to clean and turn off the tap

Jack: Excuse me, sir, what should/shall we do before cooking?

Chef: You must wash your hands. First you can wet hands with soap. Then _____.

Jack: What shall I do next?

Chef: After that you can _____ and _____.

Jack: What shall I do at last?

Chef: Finally you can _____.

Task 3   Other hygiene and safety rules in the kitchen 其他卫生安全常识

Part 1   The ways of dealing with hurt and illness. 处理外伤和疾病的正确方法。

厨师在工作中碰到外伤和疾病情况要学会正确处理，下面会告诉大家一些正确的处理方法。

Activity Read and decide true (T) or false (F). 读短文并判断正误。

Safety in the kitchen

1. Hands must be washed with water for a minimum(至少)of 10 seconds.

2. All cuts must be bandaged(缠着绷带)with waterproof band-aids and gloves should be worn.

3. Kitchen staff with sore throats(咽喉疼)or any infectious diseases(传染病)shall not be permitted to work in the kitchen.

4. No eating or drinking in the kitchen area. Do not use tobacco products in the kitchen.

1. _____ You must wash your hands with water for 15 minutes.

2. _____ You needn't bandage all cuts with waterproof band-aids and shouldn't wear gloves.

3. _____ Kitchen staff with sore throats shall be permitted to work.

4. _____ Eating, drinking and smoking are permitted in the kitchen area.

 Part 2　　Other hygiene rules in the kitchen. 其他的厨房卫生知识。

作为一名西餐实习厨师,遵守厨房的其他卫生知识也是重要的事情。

 Activity 1 Read and answer the questions. 读对话并回答问题。

Jack: Excuse me, when should I turn off the fire power?

Chef: While leaving the kitchen, you must remember that.

Jack: Do I need to clean the work station?

Chef: Yes, you must do so, and you have to clean the equipment regularly.

Jack: Oh, I see. What else should be done?

Chef: Dish sinks and surrounding areas should be cleaned.

Jack: Yes, I think it's necessary to sweep and mop the floor.

Chef: Well done.

1. What should a cook do while leaving the kitchen?

_____

2. Does a cook need to clean the work station?

_____

3. What should a cook do for the equipment?
_____

4. What else should be done?
_____

Activity 2 Write down what Jack has to do in the kitchen according to **Activity 1**. 参考 **Activity 1** 写出 Jack 如何清洁厨房卫生。

1. 擦用具_____

2. 清洁水槽_____

3. 扫地和拖地_____

4. 清洁工作灶台_____

Part 3 Making dialogues. 对话练习。

A: What shouldn't we do in the kitchen?

B: You shouldn't _____.

take food home

wear shorts

Unit 1　Kitchen Introduction

keep the kitchen door open all times　　　put a pan on the floor

Necessary words and phrases（必需词汇）

executive sous-chef[ɪgˈzekjətɪvˈsuːʃef]行政副总厨　　executive chef 行政总厨
coffee shop kitchen chef 咖啡厅厨师长　　trainee[ˌtreɪˈniː]学员
banquet[ˈbæŋkwɪt] kitchen chef 宴会厨房厨师长　　kitchen helper 帮厨
pastry[ˈpeɪstri] & bakery[ˈbeɪkəri] chef 包饼房厨师长　　demi[ˈdemi] chef 领班
Italian kitchen chef 意大利厨房厨师长　　commis[ˈkɒmi] chef 助理厨师
chef de partie[ˈdi parti]厨师主管　　main [meɪn] kitchen 主厨房

Expansion words and phrases（拓展词汇）

cold kitchen 冷菜厨房　　hot kitchen 热菜厨房
butchery [ˈbʊtʃəri]肉房　　pastry & bakery 包饼房
Italian kitchen 意大利厨房　　coffee shop kitchen 咖啡厅厨房
chef hat 厨师帽　　tie 三角巾
apron[ˈeɪprən]围裙　　plaid [plæd] pants 格子裤

Notes（重点词汇、短语）

Task 1

bring food into...　　把食物带到……
lend kitchen tools to others　　把厨具借给别人
keep the knife well away from...　　使刀远离……
clean kitchen facilities and utensils　　清洗厨房设备

Task 2

rub hands　　搓手
rinse hands　　冲手
hold water　　捧水

dry hands	擦手

Task 3

for a minimum of...	至少……
be bandaged with...	用……缠着
waterproof band-aids	防水的绷带
gloves should be worn	戴上手套
sore throats	咽喉痛
infectious diseases	传染病
be permitted to do...	被允许做……
in the kitchen area	在厨房
tobacco products	烟草产品
turn off the power	关上电源
work station	工作场所
clean the equipment regularly	经常清洁厨房设备
dish sinks and surrounding areas	水槽及周围
sweep and mop the floor	扫地和拖地

Culture life（文化生活）

Welcome to the basic work practice. 欢迎参加基础厨房工作实践

To set up a work station we need: chopping board (red, yellow, blue, brown, green, white). 摆放工作台我们需要的是：砧板（红色，黄色，蓝色，棕色，绿色，白色）。

Notes：根据各个国家和不同酒店的规则，你应该遵守不同的要求。

How to set up a work station？如何摆放工作台？

   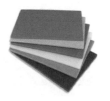

wet cloths 湿布   spoon and container 勺子和容器   knife 刀   chopping board 砧板

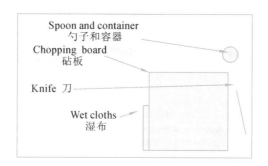

Practice 巩固练习

Exercise 1 Translate these words into Chinese or English.（中英单词互译）

1. trainee _____   2. 厨师 _____

3. chef de partie _____   4. 厨房 _____

5. mop the floor _____   6. 洗手 _____

7. pastry & bakery chef _____   8. 清洁水槽 _____

9. gas prevention _____   10. 防止火灾 _____

Exercise 2 Tick the different word.（找不同）

1. A. cold kitchen   B. executive chef   C. hot kitchen   D. butchery

2. A. wear shorts   B. fire prevention   C. electric safety   D. gas prevention

3. A. chef hat   B. towel   C. apron   D. steamer

4. A. demi chef   B. chef de partie   C. chief steward   D. trainee

5. A. gloves   B. uniforms   C. chef   D. shorts

Exercise 3 Write the sentences in the right order.（排序写句子）

1. in charge of, the pastry & bakery, I, am

_____

2. to sweep and mop, is, it, necessary, the floor

_____

3. should, 10 seconds, your hands, keep, for, washing, you

_____

4. waterproof band-aids, must be bandaged, all cuts, with

_____

5. smoke, you, to, in the kitchen, are not allowed

_____

Exercise 4 Match the key phrases with these pictures.（连线）

knife safety clean dish sink gas prevention clean a pan fire prevention

Exercise 5 Find the answers.（找答案）

1. Where do you work?　　　　　　　A. After using the toilet.

2. When should I wash my hands?　　　B. I work in Hilton Hotel.

3. What do you do in the kitchen?　　　C. Dish sink and surrounding areas should be cleaned.

4. May I smoke in the kitchen?　　　　D. No, you can't.

5. What else should be done?　　　　　E. I am a pastry and bakery chef.

Exercise 6 Talk about your favourite job.（说说你最喜爱的工作）

A: What would you like to be in the future?

B: I want to be a hot kitchen chef.

1. I want to be a _____（冷菜厨房厨师长）.

2. Mary wants to be a _____（厨师领班）.

3. We want to be a _____（咖啡厅厨师长）.

4. You want to be a _____（包饼房厨师长）.

5. They want to be _____（厨师）.

6. Jack wants to be an _____（行政总厨）.

Exercise 7 Translate the following sentences into Chinese or English.（中英文句子翻译）

1. 我在希尔顿酒店工作。

_____

2. 包饼房厨师长是负责做蛋糕的。

_____

3. 你在厨房做什么工作？我是一名实习生。

4. Dish sinks and surrounding areas should be cleaned.

5. Nobody is allowed to smoke in the kitchen.

Unit 1
参考答案

Unit 2
Kitchen Equipment

You will be able to:

1. tell the names of kitchen equipment;
2. describe the usage and function of kitchen equipment;
3. demonstrate the ability of using the equipment.

Unit introduction 单元介绍

本单元我们将学习厨房设备,主要介绍常用设备、机械设备、制冷设备和辅助设备。本单元的任务包括学习厨房设备的名称、基本功能和用途。

扫码看课件

Thinking map（思维导图）

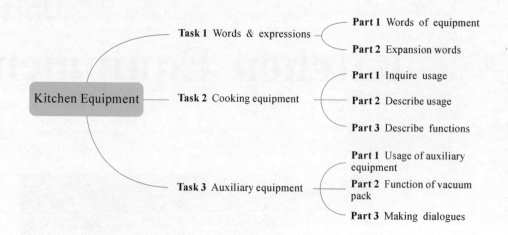

Warming up

Activity 1 Look and match. 看单词，连线。

freezer　　dishwasher　　microwave　　stove　　toaster

外教有声　　**Activity 2** Listen and tick. 听录音，勾出对话中提到的设备。

Unit 2 Kitchen Equipment

Task 1   Words & expressions 词汇

Part 1   Words of equipment. 厨房设备名称词汇。

 Activity 1 Read and write. 看图写出下列设备的英文名称。

fryer 炸炉　　　griddle 扒炉　　　oven 烤箱　　　steamer 蒸箱
salamander 焗炉　tilting boiler 汤炉　kneader 和面机　fermentation machine 醒发箱

 Activity 2 Listen and number. 听录音,标出录音中提到的设备名称的顺序。

（　）mixer 搅拌机　　（　）bone saw 骨锯　　（　）slicer 切片机
（　）mincer 绞肉机　　（　）blender 果汁机　　（　）dough pressing machine 压面机

Activity 3 Look and translate. 单词互译。

 骨锯 _____

 slicer _____

绞肉机_____   mixer _____

 压面机_____   blender _____

Activity 4 Look and write. 根据下图和描述写出设备英文名称。

1. _____ It is used to make ice.

2. _____ It is used to keep food fresh at low temperature.

3. _____ It is used to store food for a longer time without losing its flavor and nutrients.

4. _____ It is used to provide hot water in the kitchen.

5. _____ It is often used to disinfect tableware.

Part 2   Expansion words. 拓展词汇。

Activity 1 Look and match. 看图连线。

华夫饼机   冰激凌机   电动肉锤   比萨饼机   电暖汤池

ice cream machine   electric bain-marie   electric meat tenderizer   waffle iron   conveyor pizza oven

Unit 2 Kitchen Equipment

 Activity 2 Listen and write. 写出录音中内容。

_____

_____

外教有声

Activity 3 Fill in the missing letters and make a self-evaluation. 补全单词并做自我评价。

1. kn___der 2. gri___le 3. s__l__m__n__er 4. t__l__ing b___ler
5. sl___er 6. bl___der 7. m__nce__ 8. __en__eriz__r
9. f___m__n__a__ion 10. st___mer

Assessment: If you can write 8-10 words, you are perfect.

If you can write 4-7 words, you are good.

If you can write 1-3 words, you will try again.

Task 2   Cooking equipment 烹饪设备

Part 1   Inquire usage. 询问设备的使用方法。

Activity 1 Write down the equipment needed. 写出烹制下列食物所需设备的名称。

_____   _____

Activity 2 Match the picture with the right phrase. 匹配图片与短语。

1. liquidize fruits   A.

2. roast chicken   B.

3. ferment dough   C.

4. whisk cream   D.

5. grill meat   E.

6. slice meat   F.

🎧 **Activity 3** Listen and fill in the blanks. 听录音，完成对话。

| A. would you like to show me how | B. electric meat tenderizer |
| C. could you tell me how | D. what shall I do with |

Jack: Chef, _____ the beef?

Chef: Use _____ to make it tender.

Jack: Okay. _____ to use it?

Chef: Sure. Simply put the beef under the machine and turn the power on.

Jack: Wow. It works. Thanks! By the way, _____ to use the fermentation cabinet?

Chef: Sure. First set fermentation time and humidity and then put dough in.

Jack: I've learned a lot. Thank you, chef.

Chef: Sure.

Part 2   Describe usage. 描述设备的使用方法。

Activity 1 Look and describe. 看图并描述设备的使用方法。

Activity 2 Unscramble the sentences. 参考 **Activity 1** 重新排序。

1. Open the oven door and put the food in.

2. When the timer goes off, open the oven door and remove the food using oven mitts.

3. Preheat the oven to the temperature specified in the recipe.

4. Close the oven door and set a timer.

Part 3 Describe functions. 描述设备的功能。

Activity 1 Tick the equipment for making dough. 标出和面用的设备。

A.　　　　　　B.　　　　　　C.　　　　　　D.

Activity 2 Look and translate. 看图，翻译英文。

1. It's used for making dough smooth and glossy.
_____

2. It's used for cutting bones.
_____

3. It's used for mincing meat.
_____

Task 3　Auxiliary equipment 辅助设备

Part 1　Usage of auxiliary equipment. 辅助设备的用途。

Activity 1 Read the passage. 阅读下文。

In the western kitchen, the auxiliary equipment refers to the electronic equipment. It is not under the control of the central processing unit. Ice making machine, hot water boiler, dish washer, freezer, refrigerator, bain marie and disinfection cabinet are common auxiliary equipment in the western kitchen. Ice making machine is used to make ice. Hot water boiler is used to provide hot water in the kitchen. Dishwasher is used to clean kitchenware and tableware. Freezer is used to freeze food; while refrigerator is used to keep food fresh under a certain temperature. Bain marie is used to keep warm gravy ready to serve. Disinfection cabinet is usually used to disinfect tableware.

Activity 2 Decide true(T) or false(F). 判断正误。

(　　) 1. The auxiliary equipment is a part of the central processing unit.

(　　) 2. Hot water boiler is used for heating food in the kitchen.

(　　) 3. We put fruits and vegetables in the refrigerator to keep fresh.

(　　) 4. We use dishwasher to clean pots and pans in the kitchen.

(　　) 5. Chefs usually use bain marie to melt chocolate.

Activity 3 Look and write. 看图写出以下辅助设备的用途。

A　　　　　　　　　B　　　　　　　　　C

A. _____

B. _____

C. _____

Unit 2 Kitchen Equipment   25

Part 2   Function of vacuum packer. 抽真空机的功能。

Activity 1 Read the dialogue. 阅读对话。

Jack: Chef, could you tell me what's a vacuum packer used for?
Chef: It's used for storing food for a longer time without losing its flavor and nutrients.
Jack: Great. What else can it be used for?
Chef: Oh, it can also help save space. When the gas is extracted, the packaged food become smaller.
Jack: Yeah. Very true.
Chef: Besides, it is often used to prevent food from crushing while shipping.
Jack: Wow. It's really useful.
Chef: Definitely.

Activity 2 Tick the functions. 勾出对话中提到的抽真空机的功能。

☐ keep fresh   ☐ keep flavor   ☐ save space   ☐ prevent crushing
☐ avoid losing nutrients   ☐ easy shipping   ☐ easy packing

Activity 3 Write down the functions. 写出对话中抽真空机的3个功能。

_____
_____
_____

Part 3   Making dialogues. 对话练习。

A: What is this called in English, chef?
B: It is _____.
A: What's this for?
B: It is used for _____.
A: Could you tell me how to use it?
B: _____.

西餐烹饪英语

disinfection cabinet 消毒柜

hot water boiler 热水炉

refrigerator 冷藏柜

Necessary words（必需词汇）

steamer [ˈstiːmə(r)] 蒸箱
fryer [ˈfraɪə(r)] 炸炉
griddle [ˈɡrɪdl] 扒炉
slicer [ˈslaɪsə] 切片机
mincer [ˈmɪnsə(r)] 绞肉机
blender [ˈblendə(r)] 榨汁机
mixer [ˈmɪksə(r)] 搅拌机
kneader [ˈniːdə] 和面机
oven [ˈʌvn] 烤箱
nutrient [ˈnjuːtriənt] 营养物

ferment [fəˈment] 发酵
grill [ɡrɪl] 烧烤
whisk [wɪsk] 搅拌
roast [rəʊst] 烘烤
humidity [hjuːˈmɪdəti] 湿度
liquidize [ˈlɪkwɪdaɪz] 榨汁
preheat [ˌpriːˈhiːt] 预热
dough [dəʊ] 面团
salamander [ˈsæləmændə(r)] 焗炉
slice [slaɪs] 切片

Expansion words and phrases（拓展词汇）

ice cream machine 冰激凌机
waffle iron [ˈwɒfl ˈaɪən] 华夫饼机
electric bain-marie [ɪˈlektrɪk ˌbæn məˈriː] 电暖汤池
electric meat tenderizer [ˈtendəraɪzə(r)] 电动肉锤

vacuum [ˈvækjuəm] pack 抽真空机
conveyor [kənˈveɪə(r)] pizza oven 比萨烤箱

Notes（重点词汇、短语）

Task 1

tilting boiler	汤炉
bone saw	骨锯
fermentation machine	醒发箱
dough pressing machine	压面机
disinfection cabinet	消毒柜

Task 2

remove the food	取出食物
oven mitt	烤箱手套
temperature specified in the recipe	食谱规定的温度
smooth and glossy	光滑细腻的
mince meat	绞肉馅

Task 3

auxiliary equipment	辅助设备
refer to	指的是
the central processing unit	核心加工设备
keep fresh	保持新鲜
save space	节约空间
packaged food	包装食品
keep from crushing	防止压碎
easy shipping	便于运输
easy packing	便于包装

Culture life（文化生活）

厨房设备使用安全常识

使用厨房设备时要检查其是否运作正常，厨房设备要有专人操作，严格按设备操作流程进行，严禁多人同时操作；只有设备完全停止后，才能进行下一步工作。

必须定期检查厨房内的燃气燃油管道、阀门，防止泄漏。如发现燃气泄漏首先关闭阀门，及时通风，并严禁使用任何明火和启动电源开关。

工作结束后，操作人员应及时关闭所有的燃气燃油阀门，切断电源、火源。

 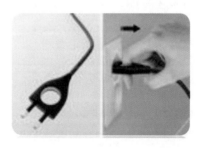

Practice 巩固练习

Exercise 1 Translate these words into Chinese or English.（中英单词互译）

1. blender _____
2. 醒发箱 _____

3. 汤炉_____   4. slicer _____
5. mincer _____   6. 焗炉 _____
7. 压面机_____   8. kneader _____
9. mixer _____   10. 扒炉_____

Exercise 2 Choose the best answer.（选择最佳答案）

1. I'd like to use _____ to make a glass of orange juice.
 A. tilting boiler   B. mincer   C. blender   D. slicer

2. We usually use _____ to clean kitchenware and tableware.
 A. dishwasher   B. stove   C. toaster   D. fryer

3. _____ is used to store food for a longer time without losing its flavor and nutrients.
 A. Freezer   B. Vacuum packer   C. Ice machine   D. Refrigerator

4. There is no _____, otherwise I can use it to color the dish.
 A. bone saw   B. salamander   C. steamer   D. fermentation machine

5. _____ is very useful, it can help keep gravy warm ready to serve.
 A. Fryer   B. Tilting boiler   C. Bain-marie   D. Steamer

Exercise 3 Match the key phrases with these pictures.（连线）

slice meat   whisk cream   grill meat   liquidize fruits

Exercise 4 Write the sentences in the right order.（排序写句子）

1. we, slicer, meat, slice, can, to, use
_____

2. a, used, vacuum pack, for, to, time, is, food, store, longer
_____

3. dough, kneader, is, making, for, used
_____

4. is, to, to, bain marie, used, keep, ready, warm, gravy, serve, until
_____

5. use, can, salamander, we, to, dish, color, the
_____

Exercise 5 Write down the equipment names.（写出设备英文名称）

1._____   2._____   3._____   4._____

5._____   6._____   7._____   8._____

Exercise 6 Find the answers.（找答案）

1. What do we call these?   A. You can use electric meat tenderizer.
2. Shall I put it into the freezer?   B. We use it to make the dough smooth and glossy.
3. How can I make the meat tender?   C. Yes, but only for 1 or 2 minutes.
4. What is dough pressing machine used for?   D. For about 20 minutes.
5. How long will you bake it?   E. They are called salamander and griddle.

Exercise 7 Translate the following sentences into Chinese or English.（中英文句子翻译）

1. 将肉汁置于电暖汤池内保温。

_____

2. 用搅拌机搅拌糖和牛奶。

_____

3. 榨汁机可以用来榨蔬菜汁和水果汁。

_____

4. Disinfection cabinet is usually used to disinfect tableware.

_____

5. Vacuum pack is often used to prevent food from crushing while shipping.

_____

Unit 3
Tools & Utensils

You will be able to:

1. know the words of different tools and utensils;
2. describe the process of using the tools;
3. talk about the process of making dishes.

Unit introduction 单元介绍

本单元我们将学习西餐刀具、炊具及厨房辅助工具相关的英语词汇和句型。本单元的学习任务包括学习工具类单词,以及用英文描述工具的功能等。

扫码看课件

Thinking map(思维导图)

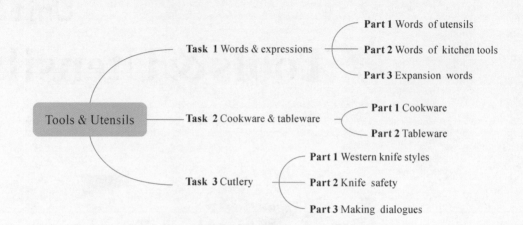

Warming up

Activity 1 Look and match. 看图和单词,连线。

| plate | bowl | fork | spoon |

外教有声

🎧 **Activity 2** Listen and tick. 听录音,勾出妈妈需要摆放的用餐工具。

Unit 3　Tools & Utensils

Task 1　Words & expressions 词汇

Part 1　Words of utensils. 厨房用具单词。

 Activity Read and write. 读单词，看图写英文。

brush 刷子　　　　　grater 擦菜板　　　　skimmer 撇沫器　　　whisk 打蛋器
flour sieve 面粉筛　　electronic scale 电子秤　rolling pin 擀面杖　chopping board 砧板

Part 2　Words of kitchen tools. 厨房工具类单词。

 Activity 1 Listen and number. 听录音，标出正确顺序。

（　）colander 沥水篮　　　（　）baking pan 烤盘　　　（　）bakery mold 蛋糕模具
（　）strainer 过滤器　　　（　）mixing bowl 搅拌碗　　（　）spatula 刮刀
（　）ladle 长柄勺　　　　　（　）measuring cup 量杯　　（　）meat tenderizer 肉锤

Activity 2 Look and translate. 单词互译。

沥水篮＿＿＿＿＿　　　　烤盘＿＿＿＿＿　　　　过滤器＿＿＿＿＿

mixing bowl _____   bakery mold _____

Activity 3 Look and write. 根据描述写出英文单词。

1. _____ A tool which makes meat softer and easier to cut.

2. _____ A large long-handled spoon with a cup-shaped bowl, which is used for serving soup.

3. _____ A metal or plastic container used for measuring quantities when cooking.

4. _____ A tool with a broad flat blade(刀片)used for mixing and spreading things in cooking.

Part 3　Expansion words. 拓展词汇。

Activity 1 Look and match. 看图连线。

油炸篮　水果切块器　防烫夹　锅盖　串肉扦　多用壶　榨汁器　铲子

tong　fruit cutter　juicer　multi-kettle　lid　frying basket　slice　skewer

🎧 **Activity 2** Listen and write. 听录音写单词。

Unit 3 Tools & Utensils

Activity 3 Fill in the missing letters and make a self-evaluation. 补全单词并做自我评价。

1. sk __ mmer   2. gr __ ter   3. fl __ __ r sieve   4. wh __ sk
5. electronic s __ ale   6. br __ sh   7. r __ lling pin   8. chopping b __ ard

Assessment: If you can write 8 words, you are perfect.
　　　　　　 If you can write 4-7 words, you are good.
　　　　　　 If you can write 1-3 words, you will try again.

Task 2　Cookware & tableware
烹饪工具和餐具

Part 1　Cookware. 烹饪工具。

Activity 1 Write down the names of cookware. 写出下面烹饪工具的名称。

_____　　　_____　　　_____　　　_____

Activity 2 Look and choose. 看图并选择相应的名称和用途。

1. braising pan		A. fry meat
2. frying pan		B. stew beef
3. stockpot		C. make sauce
4. saucepan		D. make soup

🎧 **Activity 3** Listen and fill in the blanks. 听录音，完成对话。

> A. fry meat with it
>
> B. you'd better use a saucepan
>
> C. Go ahead
>
> D. steam potatoes with a double boiler

Jack：What's this?

Chef：It's a frying pan.

Jack：Oh, What can I do with it?

Chef：You can _____.

Jack：Shall I make the sauce in a braising pan?

Chef：Well, _____.

Jack：Shall I stew beef in a braising pan?

Chef：_____.

Jack：Shall I _____?

Chef：Yes. The double boiler is in the cupboard.

Jack：All right.

Activity 4 Look and describe. 看图并描述汤锅的用途。

_____   _____   _____   _____

🎧 **Activity 5** Listen and complete the dialogue. 听录音，参考 **Activity 3** 完成对话。

> A. Strain broth through a sieve
>
> B. Put it in the stockpot
>
> C. Then lower the fire
>
> D. Bring the water to boil

Jack：Where shall I put the beef?

Chef：_____.

Jack：What's next?

Chef：_____. Don't fill it to the top.

Jack: And then?

Chef: _____. Stew until almost out of water.

(2 hours later...)

Jack: To strain?

Chef: _____.

Part 2   Tableware. 餐具。

Activity 1 Write down the correct names of drinkware. 写出饮具的正确名称。

1. Drink tea with _____

2. Drink coffee with _____

3. Drink red wine with _____

4. Drink white wine with _____

5. Drink juice with _____

6. Eat ice cream with _____

Activity 2 Make dialogues with the dishware. 根据所给餐具进行对话练习。

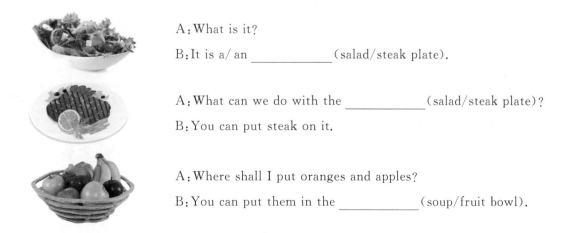

A: What is it?
B: It is a/an _____ (salad/steak plate).

A: What can we do with the _____ (salad/steak plate)?
B: You can put steak on it.

A: Where shall I put oranges and apples?
B: You can put them in the _____ (soup/fruit bowl).

A: What size is the _____ (soup/ fruit bowl)?
B: It is _____ (7.5/ 9 inches).

Task 3　Cutlery 刀具

> 在厨房里，刀具各种各样，是厨师们的左膀右臂，专为完成不同的任务而设计。它们可以帮助厨师完成对各种食材的分割。

Part 1　Western knife styles. 西餐刀具的种类。

 Activity 1 Look and read. 读一读。

Chef: You know Jack, kitchen knives are intended to be used in food preparation. They are designed for specific tasks.

Jack: Which tool is commonly used in the kitchen?

Chef: Chef's knife is most commonly used for a chef. It can be used to slice, shred and chop.

Jack: Wonderful. What are these tools for?

Chef: Paring knife is used for cutting fruits. Spatula is the tool used for mixing and spreading things. Bread knife is used for cutting bread. We can cut cheese with a cheese knife; chop bones with a boning knife; cut salmon with a salmon knife and open oysters with an oyster knife.

Jack: They are very useful, right? This tool looks very special, what to do with it?

Chef: It is called fish scale, you can scale fish with it. Besides, kitchen scissors are also useful for cutting smaller fish. Remember, they have to be kept sharp!

Jack: What else can I use to sharpen my knife?

Chef: You may use a whetstone or sharpening steel. You can also use an electric knife sharpener. It is safer.

Jack: Oh, I see. Thank you for telling me so much.

Activity 2 Write down the tools of sharpening a knife. 参考 **Activity 1** 写出磨刀工具的英文形式。

_____   _____   _____

Activity 3 Match the picture with the right phrase. 连线。

1. bread knife      A. slice bread

2. boning knife      B. bone chicken legs

3. chef's knife      C. slice salmon

4. oyster knife      D. cut up vegetables

5. salmon knife      E. open oysters

Part 2 Knife safety. 刀具安全。

厨师刀能帮助厨师们解决工作中的困难,同时增添工作乐趣。

Activity 1 Read the passage. 阅读短文。

Here are some tips for staying safe while using knives. Remember them!

Do

√ 1. Use a knife suitable for the task and the food you are cutting.

√ 2. Keep knives sharp.

√ 3. Handle knives carefully when washing up.

√ 4. Cut on a stable surface.

√ 5. Train employees to use knives safely when sharpening.

Don't

× 1. Try to catch a falling knife.

× 2. Carry a knife in your pocket.

× 3. Use a knife as a can opener.

× 4. Carry knives while carrying other objects.

× 5. Leave knives loose on worktop surfaces where they can be accidentally pushed off.

Activity 2 Decide true(T) or false(F). 判断正误。

(   ) 1. When a knife falls down, you may catch it with your hands.

(   ) 2. You may use a knife as a can opener.

(   ) 3. Please cut vegetables on a stable surface when you prepare food.

(   ) 4. You will always keep your knives sharp.

(   ) 5. You may carry knives while carrying other objects.

Part 3   Making dialogues. 对话练习。

Example 1

A: What are you going to do?

B: I'm going to scale fish with a fish scale.

cut fruit/ paring knife

slice potatoes/ chef's knife

cut cheese/ cheese knife

open fish/ kitchen scissors

Example 2

A: What can't we do?

B: We can't _____.

Unit 3 Tools & Utensils

catch a falling knife

use a knife as a can opener

carry a knife in your pocket

Necessary words and phrases（必需词汇）

brush [brʌʃ] 刷子

braising pan [breɪz pæn] 炖锅

double boiler ['dʌbl 'bɔɪlə(r)] 双层蒸锅

baking pan ['beɪkɪŋ pæn] 烤盘

saucepan ['sɔːspən] 长柄锅

spatula ['spætʃələ] 刮刀

boning knife ['bəʊnɪŋ naɪf] 剔骨刀

paring knife ['peərɪŋ naɪf] 水果刀

oyster knife ['ɔɪstə(r) naɪf] 牡蛎刀

grater ['greɪtə(r)] 擦菜板

whisk [wɪsk] 打蛋器

ladle ['leɪdl] 长柄勺

mixing bowl ['mɪksɪŋ bəʊl] 搅拌碗

stockpot ['stɒkpɒt] 汤锅

bread knife [bred naɪf] 面包刀

salmon knife ['sæmən naɪf] 三文鱼刀

chef's knife [ʃefs naɪf] 主厨刀

peeler ['piːlə(r)] 削皮刀

外教有声

Expansion words and phrases（拓展词汇）

skimmer ['skɪmə] 撇沫器

strainer ['streɪnə(r)] 过滤器

flour sieve ['flaʊə(r) sɪv] 面粉筛

lid [lɪd] 锅盖

colander ['kʌləndə(r)] 沥水篮

tongs [tɒŋz] 防烫夹

frying basket ['fraɪɪŋ 'bɑːskɪt] 油炸篮

juicer ['dʒuːsə(r)] 榨汁机

bakery mold ['beɪkəri məʊld] 蛋糕模具

rolling pin ['rəʊlɪŋ pɪn] 擀面杖

fruit cutter [fruːt 'kʌtə(r)] 水果切块机

meat tenderizer [miːt 'tendəraɪzə] 肉锤

chopping board ['tʃɒpɪŋ bɔːd] 砧板台

measuring cup ['meʒərɪŋ kʌp] 量杯

electronic scale [ɪˌlek'trɒnɪk skeɪl] 电子秤

multi-kettle ['mʌlti 'ketl] 多用壶

外教有声

Notes（重点词汇、短语）

Task 1

long-handled   长柄的

cup-shaped	杯状的
metal	金属的
plastic	塑料的
measuring quantities	测量数量
a broad flat blade	宽大、扁平的刀片
spreading	铺展

Task 2

stew	炖
steam	蒸
cupboard	橱柜
strain broth	过滤汤汁
lower the fire	把火调小
simmer	小火煨

Task 3

commonly	普遍地
intend to	为了
specific	特别的
a stable surface	平稳的表面
suitable for	适用于
handle knives	握刀

Culture life（文化生活）

Table setting 西餐摆台

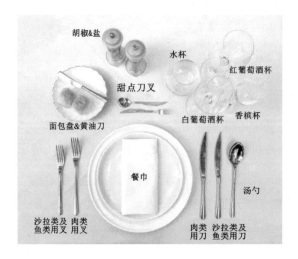

Unit 3　Tools & Utensils

①黄油刀 butter knife

②点心匙及点心叉 dessert spoon and cake fork

③水杯 sterling water goblet

④红酒杯 red wine goblet

⑤白酒杯 white wine goblet

⑥鱼叉 fish fork

⑦主菜叉 dinner or main course fork

⑧主菜盘 dinner plate

⑨餐巾 napkin

⑩主菜刀 dinner knife

⑪鱼刀 fish knife

⑫汤匙 soup spoon

Practice 巩固练习

Exercise 1 Translate these words into Chinese or English.（中英单词互译）

1. chef's knife _____   2. 比萨刀 _____
3. cheese knife _____   4. 刮刀 _____
5. mixing _____   6. 切片 _____
7. spreading _____   8. 切成末 _____
9. whetstone _____   10. 切成丝 _____

Exercise 2 Tick the different word.（找不同）

1. A. bread knife   B. pizza knife   C. cheese knife   D. small knife
2. A. whisk   B. ice cream   C. grater   D. rolling pin
3. A. double boiler   B. frying pan   C. saucepan   D. kettle
4. A. slice   B. shred   C. small   D. chop
5. A. spatula   B. chef's knife   C. paring knife   D. sharpen

Exercise 3 Match the key phrases with these pictures.（连线）

mince the garlic　　slice the meat　　shred the cabbage　　dice the carrot

Exercise 4 Write the sentences in the right order.（排序写句子）

1. last, think, is, to, thing, for, plating, the, about
_____

2. like, is, a chef's knife, what
_____

3. soup, for, special, the, bowl, is
_____

4. the, called, tool, be, must, sauté pan
_____

5. do, not, forget, to, coffee, serve, with, coffee cup
_____

6. mix, we, with, a whisk, eggs
_____

Exercise 5 Fill in the blanks with the proper words.（选择正确的选项）

1. Sift the flour with a _____   A. grater
2. Drain the carrots in the _____   B. whisk
3. Grate the potato with a _____   C. sieve
4. Fry the egg in the _____   D. frying pan
5. Mix the egg with a _____   E. colander

1. _____ the pan off the fire.   A. Boil
2. _____ the meat in the hot pot.   B. Take
3. _____ the plate with the tongs.   C. Chop
4. _____ the surface of the oil with a soup ladle.   D. Lift
5. _____ the potato on the chopping board.   E. Clean

Exercise 6 Translate the following sentences into Chinese or English.（中英文句子翻译）

1. We serve salad and fruit with deep plates.
_____

2. What is this special knife for?
_____

3. 厨师刀用途广泛，可以用来切菜、切肉等。
_____

4. 请把牛排装在牛排盘子里。
_____

5. 我们用煎锅来煎蛋。
_____

Unit 4
Condiments & Spices

You will be able to:

1. know the words of condiments and spices;
2. classify condiments and spices;
3. talk about the use of condiments.

Unit introduction 单元介绍

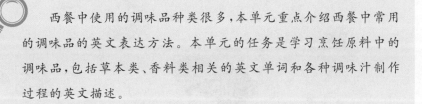

西餐中使用的调味品种类很多,本单元重点介绍西餐中常用的调味品的英文表达方法。本单元的任务是学习烹饪原料中的调味品,包括草本类、香料类相关的英文单词和各种调味汁制作过程的英文描述。

扫码看课件

西餐烹饪英语

Thinking map（思维导图）

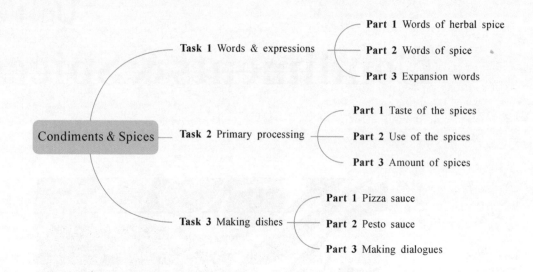

Warming up

Activity 1 Look and match. 看单词，连线。

| mustard | ketchup | honey | vinegar |

外教有声

Activity 2 Listen and tick. 听录音，勾出女孩在超市购买的调味品。

Note

Task 1 　 Words & expressions 词汇

Part 1 　 Words of herbal spice. 草本调料单词。

 Activity Read and write. 读单词，看图写英文。

| oregano 牛至 | thyme 百里香 | dill 莳萝 | parsley 欧芹 |
| rosemary 迷迭香 | basil 罗勒 | mint 薄荷 | chive 细叶葱 |

_____　　_____　　_____　　_____

_____　　_____　　_____　　_____

Part 2 　 Words of spice. 香料单词。

 Activity 1 Listen and number. 听录音，标出正确顺序。

( ) bay-leaf 香叶　　　( ) clove 丁香　　　( ) paprika 红辣椒粉
( ) curry powder 咖喱粉　　　( ) cinnamon 肉桂　　　( ) mustard 芥末
( ) cardamom 小豆蔻　　　( ) nutmeg 肉豆蔻

Activity 2 Look and translate. 单词互译。

香叶_____　　咖喱粉_____　　cinnamon_____　　paprika_____

Activity 3 Look and write. 根据描述写出单词。

1. _____ Dried leaves of the bay laurel（月桂树）.
2. _____ A mild powdered seasoning which is made from red pepper.
3. _____ The leaves of a mint plant used fresh or candied（甜蜜的）.
4. _____ Leaves of the common basil used fresh or dried.

Part 3   Expansion words. 拓展词汇。

Activity 1 Look and match. 看图连线。

藏红花   姜黄   鼠尾草   莳萝籽   芫荽   墨角兰   柠檬草   龙蒿

cumin   sage   coriander   lemon grass   tarragon   saffron   turmeric   marjoram

Activity 2 Listen and write. 听录音写单词。

_____

_____

Activity 3 Fill in the missing letters and make a self-evaluation. 补全单词并做自我评价。

1. p __ __ sley   2. b __ sil   3. p __ prika   4. th __ me   5. bay-l __ __ f

6. must __ d   7. cl __ve   8. n __tmeg   9. __ __egano   10. ch __ve

Assessment: If you can write 8-10 words, you are perfect.

If you can write 4-7 words, you are good.

If you can write 1-3 words, you will try again.

Task 2 Primary processing 初加工

Part 1   Taste of the spices. 调料的味道。

Activity 1 Write down the different tastes. 写出不同的味道。

_____   _____   _____

sweet	bitter
sour	salty
hot	

_____   _____

Activity 2 Write and match. 写出单词并按调料口味分类连线。

盐   savory   sweet   果酱

蜂蜜   桂皮   红辣椒粉   香草   醋   黑胡椒

🎧 **Activity 3** Listen and fill in the blanks. 听录音，完成对话。

A. What should we make today?   B. put some honey

C. make cheese cake   D. add some other spices

外教有声

Jack: Chef? _____

Chef: Let's make a cake.

Jack: I like chocolate cake very much.

Chef: We will _____. It isn't very sweet.

Jack: Can we put less sugar in it?

Chef: If you like, you can _____ instead of sugar.

Jack: Really?

Chef: It will make the taste better.

Jack: Do I need to _____?

Chef: Vanilla is also a useful spice for a bakery.

Jack: I know. Thank you.

Part 2 Use of the spices. 调料的使用。

Activity 1 Look and write. 看图并写出这些调料的名称。

_____   _____   _____   _____

Activity 2 Listen and complete the dialogue. 听录音，完成对话。

| A. add some wild pepper | B. to color the soup and dish |
| C. is widely used in soup and sauce | D. in pickles |

Jack: Could you tell me how to use oregano and mustard, chef?

Chef: Oregano _____, especially when we make vegetable soup.

Jack: How about fennel, cinnamon and bay-leaf?

Chef: They are often used _____.

Jack: Is it important to _____ and chili to make spicy food?

Chef: Yes. But paprika is a special one, it is not hot, it is only used _____.

Jack: Thanks a lot. I get it.

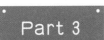

Unit 4　Condiments & Spices

Part 3　Amount of spices. 调料的量。

 Activity 1 Read and match. 量词的缩写形式及中文翻译连线。

外教有声

teaspoon	g	汤匙
gram	L	毫升
tablespoon	oz	磅
gallon	mL	茶匙
liter	gal	盎司
pound	tsp	克
ounce	tbs	升
milliliter	lb	加仑

Activity 2 Look and translate. 看图，翻译英文。

1. People need to drink at least 2000 mL water every day.
   _____

2. Help me to get a teaspoon of sugar, please.
   _____

3. You need to add two tablespoons of mustard.
   _____

4. Americans use gallons to show the amount of milk.
   _____

Task 3　Making dishes 制作菜肴

Part 1　Pizza sauce. 比萨酱。

比萨酱的口味很基础，不突出任何特殊风味，所以使用范围非常广泛，也可以在此基础上添加其他食材而衍生出其他常吃的酱料。

Activity 1 Read the recipe. 读菜单。

① Ingredients

tomato
ketchup
onion
olive oil
garlic
basil
bay-leaf
salt
sugar
black pepper

② Methods

1. Fry onions, chopped garlic in olive oil.
2. Put diced tomatoes in and stir until it gets soft.
3. Add some ketchup.
4. Put basil, salt, sugar, black pepper in and add a piece of bay-leaf.
5. Cover and simmer for about 20 minutes.
6. Remove the bay-leaf.

Activity 2 Tick the ingredients. 勾出所需原料。

☐ black pepper　　☐ butter　　☐ onion　　☐ ketchup　　☐ starch
☐ sesame seeds　　☐ salad oil　　☐ fennel　　☐ garlic　　☐ sugar

☐ salt　　　　☐ celery　　　☐ basil　　　☐ bay-leaf　　　☐ tomato

Activity 3 Write the processing steps according to **Activity 1**. 根据 **Activity 1** 写出制作流程。

　　　A　　　　　　　　　B　　　　　　　　　C

A. _____

B. _____

C. _____

Part 2　　Pesto sauce. 青酱（罗勒松子酱）。

> 青酱是一种冷拌酱，健康又美味，尤其受女生喜爱，大部分时候用来拌面和抹面包，现在也可以作为肉类的腌酱和蘸酱。

Activity 1 Read the dialogue. 读对话。

Jack: I will make pesto sauce. What do I need?

Chef: You should prepare the tools first. What do you think of that?

Jack: Food processor and oven are necessary.

Chef: Wonderful. Don't forget the colander.

Jack: Oh, I almost forget it. How about the ingredients?

Chef: Basil, nuts, almond, olive oil, garlic, cheese, salt and black pepper.

Jack: I've finished them, chef.

Chef: Now let's make the sauce.

Activity 2 Tool preparation. 工具准备。

沥水篮_____　　料理机_____　　烤箱_____

外教有声

Activity 3 Ingredient preparation. 原料准备。

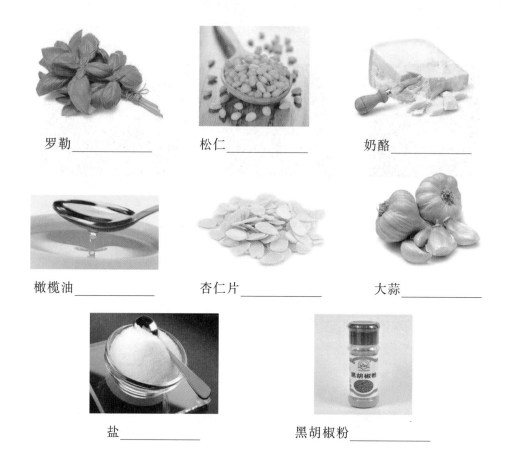

罗勒_____   松仁_____   奶酪_____

橄榄油_____   杏仁片_____   大蒜_____

盐_____   黑胡椒粉_____

Activity 4 Look and write the production process. 看图并写出制作流程。

1. Remove the pine nut shells and roast the almonds in the oven for 5-10 minutes with 180 ℃.
2. Put the basil in the colander after washing, only retain the basil leaves, then peel garlic.

3. Put basil, garlic and almonds into the cooking machine in food processor.
4. Blend them with slow speed and then add salt and pine nuts.
5. Continue to blend and add the sliced cheese, black pepper and olive oil.
6. After blending all the things fully. Put it in a clean bottle and seal it tightly.

_____   _____   _____

Unit 4　Condiments & Spices

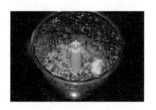
_____

_____

_____

Part 3　Making dialogues. 对话练习。

A: Can I take your order?
B: OK. I want _____.
A: What kind of sauce do you need?
B: I like _____.

vegetable salad　vinegar dressing

pasta　basil pesto sauce

French fries　tomato ketchup

Necessary words and phrases（必需词汇）

oregano [ˌɒrɪˈɡɑːnəʊ] 牛至
thyme [taɪm] 百里香
dill [dɪl] 莳萝
parsley [ˈpɑːsli] 欧芹
rosemary [ˈrəʊzməri] 迷迭香
basil [ˈbæzl] 罗勒
mint [mɪnt] 薄荷
chives [tʃaɪvz] 细叶葱

cardamom [ˈkɑːdəməm] 小豆蔻
cinnamon [ˈsɪnəmən] 肉桂
mustard [ˈmʌstəd] 芥末
nutmeg [ˈnʌtmeɡ] 肉豆蔻
mill [mɪl] 磨粉机
grinder [ˈɡraɪndə(r)] 研磨器
marjoram [ˈmɑːdʒərəm] 墨角兰
black pepper [blæk ˈpepə(r)] 黑胡椒

外教有声

curry powder ['kʌri 'paʊdə(r)] 咖喱粉　　paprika ['pæprɪkə] 红辣椒粉
bay-leaf [beɪ liːf] 月桂叶　　　　clove [kləʊv] 丁香

Expansion words（拓展词汇）

sage [seɪdʒ] 鼠尾草　　　　　　cayenne [keɪ'en] 红辣椒
cumin ['kʌmɪn] 莳萝籽　　　　　tarragon ['tærəgən] 龙蒿
coriander [ˌkɒri'ændə(r)] 芫荽　turmeric ['tɜːmərɪk] 姜黄粉
fennel ['fenl] 小茴香　　　　　saffron ['sæfrən] 藏红花
lemon grass ['lemən ɡrɑːs] 柠檬草

Notes（重点词汇、短语）

Task 1

instead of	代替
for bakery	用于烘焙店

Task 2

how about...	……怎么样
It is important to...	……是重要的

Task 3

stir	搅拌
food processor	料理机
blend	混合，掺杂

Culture life（文化生活）

不同风味的牛排

牛排，或称牛扒，块状的牛肉，是西餐中最常见的食物之一。牛排的烹调方法以煎和烧烤为主。牛排的种类非常多，常见的有 Tenderloin（菲力牛排）、Rib-eye（肉眼牛排）、Sirloin（西冷牛排/沙朗牛排）、T-bone steak（T骨牛排）和 Dry aged steak（干式熟成牛排）。

1. 加拿大风味

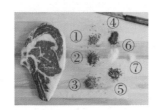
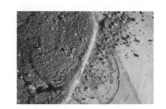

①芫荽子(crushed coriander);②黑胡椒(black pepper);③莳萝(dill);④红椒粉(paprika);⑤红辣椒片(red pepper flakes);⑥洋葱粒(granulated onion);⑦蒜粒(granulated garlic)。

2. 西班牙风味

①橙皮屑(orange zest);②烟熏辣椒粉(smoked paprika);③平叶欧芹(flat leaf parsley)。

3. 墨西哥风味

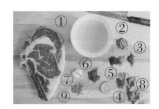 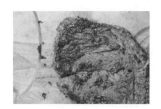

①墨西哥黑啤酒(dark Mexican beer);②小茴香(cumin);③黑胡椒(black pepper);④青柠(lime);⑤辣椒粉(chill powder);⑥干辣椒粉(dried ancho pepper);⑦蒜(garlic);⑧芫荽(coriander);⑨干墨西哥牛至(dried Mexican oregano)。

Practice 巩固练习

Exercise 1 Translate these words into Chinese or English.（中英文单词互译）

1. parsley _____   2. 红椒粉 _____
3. basil powder _____ 4. 黄芥末 _____
5. turmeric _____   6. 香叶 _____
7. crushed cumin _____ 8. 细叶葱 _____
9. oregano _____   10. 桂皮 _____

Exercise 2 Tick the different word.（找不同）

1. A. chives   B. chicken   C. mint   D. dill
2. A. fish   B. crab   C. garlic   D. shrimp
3. A. sweet   B. sour   C. hot   D. beautiful
4. A. grinder   B. curry powder C. black pepper D. oregano
5. A. parsley   B. salt   C. onion   D. mustard

Exercise 3 Match the word with these pictures.（连线）

fennel   mint   saffron   sage   chives   nutmeg

Exercise 4 Write the sentences in the right order.（排序写句子）

1. five, are, there, condiments, of, types

2. of, what, the, flavor, is, it

3. sugar, sweet, is, very

4. is, cooking, often, in, used, soup, meat and vegetable, sage

5. in, shall, I, call, it, English, what

Exercise 5 Find the answers.（找答案）

1. Would you like the coffee with sugar?   A. In a clean bottle and seal it.
2. How long should I simmer the pizza sauce?   B. Yes. Just a teaspoon of sugar, please.
3. Do you use butter to make pizza sauce?   C. I use it to strain the vegetable.
4. What do you use colander for?   D. For about 20 minutes.
5. Where should I put the pesto sauce?   E. No, I only use olive oil.

Exercise 6 Translate the following sentences into Chinese or English.（中英文句子翻译）

1. Heat the butter, honey, mustard and curry powder. Stir until butter is melted in a pot.

2. Other common herbs and spices include basil, thyme and fennel seed.

3. 我喜欢蘸着番茄酱吃薯条。

4. 将橄榄油、切碎的蒜瓣和一些黑胡椒粉放入深平底锅里加热。

5. 请用研磨器再多磨出一些黑胡椒粉。

Unit 5
Vegetables

You will be able to:

1. read and write the words of different vegetables;
2. describe the ingredients of dishes;
3. talk about the process of making vegetable dishes.

Unit introduction 单元介绍

本单元我们将学习烹饪原料中的各种蔬菜。蔬菜主要分为叶类和根茎类等。本单元的任务是学习各种蔬菜单词、初加工过程和以蔬菜为主料的典型菜肴的相关英文词汇和表达方式。

扫码看课件

西餐烹饪英语

Thinking map（思维导图）

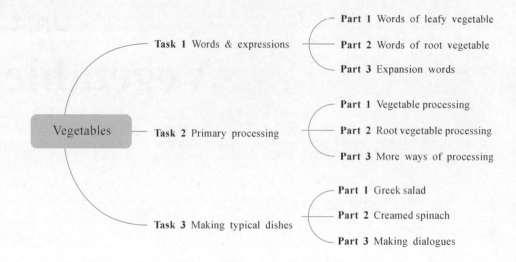

Warming up

Activity 1 Look and match. 看单词，连线。

onion　　　bean　　　celery　　　carrot

外教有声

Activity 2 Listen and tick. 听录音，勾出女孩吃的沙拉中所包含的蔬菜。

Note

Unit 5 Vegetables

Task 1 Words & expressions 词汇

Part 1 Words of leafy vegetable. 叶类蔬菜单词。

 Activity Read and write. 读单词，看图写英文。

| celery 芹菜 | lettuce 生菜 | cabbage 卷心菜 | parsley 欧芹 |
| spinach 菠菜 | spring onion 葱 | lollo rossa 红生菜 | frisée 苦苣菜 |

Part 2 Words of root vegetable. 根茎类蔬菜单词。

 Activity 1 Listen and number. 听录音，标出正确顺序。

( ) eggplant 茄子   ( ) cauliflower 花椰菜   ( ) artichoke 洋蓟   ( ) beetroot 甜菜
( ) pumpkin 南瓜   ( ) broccoli 西蓝花   ( ) asparagus 芦笋   ( ) kohlrabi 球茎甘蓝

Activity 2 Look and translate. 单词互译。

西蓝花_____   球茎甘蓝_____   洋蓟_____

西餐烹饪英语

cauliflower _____   pumpkin _____

Activity 3 Look and write. 根据描述写出英文单词。

1. _____ A long vegetable with purple color.
2. _____ A vegetable that is long and green and has small shoots at one end.
3. _____ It's a kind of vegetable with thick round and sweet root.

Part 3　Expansion words. 拓展词汇。

Activity 1 Look and match. 看图连线。

蚕豆　　红辣椒　　橄榄　　大蒜　　青椒　　西葫芦　　蘑菇　　豌豆

mushroom　garlic　chili　green pepper　zucchini　olive　broad bean　pea

Unit 5   Vegetables

 Activity 2 Listen and write. 听录音写单词。

_____

_____

Activity 3 Fill in the missing letters and make a self-evaluation. 补全单词并做自我评价。

1. z__cchin__   2. __l__ve   3. p____   4. br__ad b__an   5. ch__li
6. m__shroom   7. __ggpl__nt   8. green p__pper   9. p__mpk__n   10. br__ccol__

Assessment: If you can write 8-10 words, you are perfect.
　　　　　　 If you can write 4-7 words, you are good.
　　　　　　 If you can write 1-3 words, you will try again.

Task 2   Primary processing 初加工

Part 1   Vegetable processing. 蔬菜的加工。

Activity 1 Write down the tools for the vegetable processing. 写出此处蔬菜加工所用主要工具的名称。

_____　　　　　　　_____

Activity 2 Look and choose. 看图并选择。

　　A. Cut the onion in half　　　　B. Grind the garlic

　　C. Cut the roots of the onion　　D. Dice the onion

　　E. Slice the carrot

64 　西餐烹饪英语

1. _____　　　2. _____　　　3. _____

4. _____　　　5. _____

外教有声

Activity 3 Listen and fill in the blanks. 听录音，完成对话。

| A. slice them finely | B. dice the celery |
| C. cut onions and carrots | D. make vegetable soup |

Jack: What are we going to do?

Chef: We are going to _____.

Jack: Shall I _____.

Chef: Yes, and _____.

Jack: And then _____?

Chef: No, grind the garlic and chop the mushroom on the chopping board.

Jack: OK, I see.

Part 2　Root vegetable processing. 根茎类蔬菜的加工。

Activity 1 Look at the processing. 看黄瓜的加工过程。

Activity 2 Listen and complete the dialogue. 听录音，参考 **Activity 1** 完成对话。

A. dice the batons B. cut off the bottom

C. cut it into batons D. flatten the bottom

Jack: What shall I do with the cucumber, chef?

Chef: You should _____ first.

Jack: OK, I have finished. And then?

Chef: Second, _____.

Jack: How many are there in a baton?

Chef: Four.

Mike: Right. What next?

Chef: _____. Last, _____.

Jack: Yes, chef.

Part 3 More ways of processing. 更多的加工方法。

Activity 1 Look and translate. 翻译蔬菜初加工的短语。

1. Cut the potato to be balls.
_____

2. Scoop the cucumber into olives.
_____

3. Scoop the potato into oval balls.
_____

4. Slice the potato.
_____

5. Peel the cucumber.
_____

Activity 2 Make dialogues. 根据所给信息进行对话练习。

A: Will you _____? (crinkle potatoes)
B: Yes, I will.

A: What should I do for the next step?
B: You should _____. (match stick potatoes)

A: What're you doing now?
B: I'm _____. (cut... straw potatoes)

A: What shall I do with it?
B: You can _____. (straight potatoes)

Task 3　Making typical dishes
特色菜肴制作

Part 1　Greek salad. 希腊沙拉。

在希腊，沙拉是人们日常必不可少的一道菜，五彩斑斓的新鲜蔬菜看起来非常诱人。多吃蔬菜有益健康，这也是希腊人健康长寿的秘密之一。希腊沙拉的味道与一般调料做出的沙拉有所不同，颇有一番异域美食的感觉。

Unit 5 Vegetables

Activity 1 Read the recipe. 读菜单。

① Ingredients

onion
lemon juice
tomato
cucumber
lettuce
olive
Feta cheese(菲达奶酪)
vinaigrette(醋油沙司)

② Methods

1. Soak the onion slices in lemon juice.

2. Mix the onion and lemon juice in a large bowl and place for 10 minutes. Drain and give up (放弃，倒掉) the liquid.

3. Mix the vegetables together, add Feta cheese and lemon juice to the onion.

4. Mix with Feta cheese gently.

5. Pour the vinaigrette(醋油沙司) over the vegetables and mix gently to coat.

Activity 2 Tick the ingredients. 勾出所需原料。

☐ onion ☐ vinaigrette ☐ olive ☐ juice
☐ lettuce ☐ Feta cheese ☐ cucumber ☐ lemon

Activity 3 Write the processing steps according to **Activity 1**. 根据 **Activity 1** 写出主要制作流程。

A

B

C

A. _____

B. _____

C. _____

Part 2　Creamed spinach. 奶油菠菜。

> 菠菜中含有丰富的铁、维生素C、胡萝卜素、维生素A、维生素B_2和抗氧化剂等,营养丰富,与奶油一起食用味道更佳。

Activity 1 Read the dialogue. 读对话。

Jack: I want to make creamed spinach. What should I prepare?

Chef: Tools first. Can you tell me what tools are necessary?

Jack: A large pot, a colander, a small saucepan, a measuring cup and a spoon.

Chef: Wonderful. Don't forget the hot mitts(隔热手套).

Jack: Oh, I almost forget it. How about the ingredients?

Chef: Spinach, butter, chopped onion, minced garlic, flour, grated Parmesan cheese(帕尔玛干酪), grated nutmeg(肉豆蔻), dry mustard(干芥末), chili, salt and ground black pepper(黑胡椒碎).

Jack: I've prepared them well, chef.

Chef: Now let's make the dish.

Activity 2 Tool preparation. 工具准备。

炖锅_____　　沙司锅_____　　大勺_____

滤锅_____　　量杯_____　　隔热手套_____

Unit 5 Vegetables

Activity 3 Ingredients preparation. 原料准备。

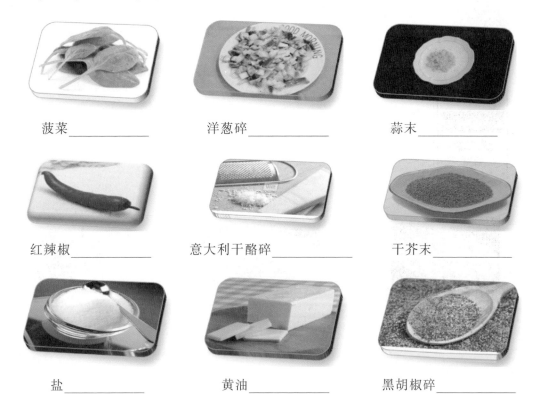

菠菜_____   洋葱碎_____   蒜末_____

红辣椒_____   意大利干酪碎_____   干芥末_____

盐_____   黄油_____   黑胡椒碎_____

Activity 4 Write the production process. 根据提示写出加工过程。

1. Add the spinach into boiling water.
2. Drain the pot of spinach in a colander.
3. Melt the butter in a frying pan over medium-high heat. Add the onion, garlic and flour.
4. Add the cheese and spinach.
5. Mix well. Keep warm over very low heat.

_____   _____   _____

   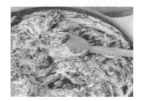

_____   _____

Part 3 Making dialogues. 对话练习。

A: What is our special today, chef?
B: It is _____.
A: How do we call it in Chinese?
B: It is called _____.

classic green bean casserole
经典青豆砂锅

minestrone soup
意大利蔬菜汤

French corn soup
法式玉米浓汤

Necessary words and phrases（必需词汇）

onion [ˈʌnjən] 洋葱
bean [biːn] 青豆
celery [ˈseləri] 芹菜
carrot [ˈkærət] 胡萝卜
lettuce [ˈletɪs] 生菜
cabbage [ˈkæbɪdʒ] 卷心菜
parsley [ˈpɑːsli] 欧芹
spinach [ˈspɪnɪtʃ] 菠菜
mushroom [ˈmʌʃrʊm] 蘑菇
garlic [ˈɡɑːlɪk] 大蒜

spring onion [sprɪŋ ˈʌnjən] 葱
eggplant [ˈeɡplɑːnt] 茄子
cauliflower [ˈkɒliflaʊə(r)] 花椰菜
artichoke [ˈɑːtɪtʃəʊk] 洋蓟
pumpkin [ˈpʌmpkɪn] 南瓜
broccoli [ˈbrɒkəli] 西蓝花
zucchini [zʊˈkiːni] 西葫芦
cucumber [ˈkjuːkʌmbə(r)] 黄瓜
green pepper [ɡriːn ˈpepə(r)] 青椒
chili [ˈtʃɪli] 红辣椒

Expansion words and phrases（拓展词汇）

lollo rossa 红生菜
frisée [friˈzeɪ] 苦苣菜
beetroot [ˈbiːtruːt] 甜菜
palm heart [pɑːm hɑːt] 棕榈心
asparagus [əˈspærəɡəs] 芦笋

broad bean 蚕豆
Feta cheese 菲达奶酪
Parmesan cheese 帕尔玛干酪
baton [ˈbætɒn] 长条，段
kohlrabi [ˌkəʊlˈrɑːbi] 球茎甘蓝

Notes（重点词汇、短语）

Task 1

| at one end | 在一端 |
| hot mitts | 隔热手套 |

Task 2

cut... in half	切成两半
slice finely	细细地切
cut off the bottom	切掉底部
cut it into batons	切成段
flatten the bottom	弄平底部
scoop... into ball	挖球
cut... into olives	切成橄榄球形
scoop... to be oval balls	挖成椭圆形
match stick potatoes	薯棍
straight potatoes	直身薯条
potato chips	薯片
straw potatoes	细薯丝
crinkle potatoes	波浪薯条

Task 3

measuring cup	量杯
grated nutmeg	肉豆蔻碎
medium-high heat	中高火
keep warm	保温
low heat	小火

Culture life（文化生活）

西蓝花清洗方法

西蓝花的花球很容易沾染上农药，需要彻底清洗才能去除农药残留。具体步骤如下。

1.把整颗的西蓝花完整放进流动的水里浸泡 10～20 分钟，以去除其表面的粉尘，再切下西蓝花可食用的部分。

2.锅里倒入足量的水，再放进西蓝花，并挤入约半颗量的新鲜柠檬。柠檬汁中的柠檬酸根带负电荷，和带正电荷的重金属中和反应，即可去除菜中的重金属残留。

3.再加入 1 小匙食用小苏打粉，搅拌均匀后浸泡约 10 分钟。小苏打粉有助于增加水的穿透性,这样可带走农药。而且如果菜里有虫,浸泡后虫子也会自动脱离。

4. 再把西蓝花放于流动的清水下彻底冲洗干净即可。西蓝花的外皮不必去除,如果仅是觉得口感太老,可只去除一小部分,去除太多会造成纤维流失。

Practice 巩固练习

Exercise 1 Translate these words into Chinese or English.(中英单词互译)

1. pumpkin _____   2. 西蓝花 _____
3. artichoke _____ 4. 青椒 _____
5. eggplant _____   6. 卷心菜 _____
7. asparagus _____ 8. 黄瓜 _____
9. lettuce _____   10. 胡萝卜 _____

Exercise 2 Tick the different word.(找不同)

1. A. eggplant   B. potato   C. onion   D. lettuce
2. A. bean   B. pea   C. broad bean   D. parsley
3. A. pumpkin   B. cabbage   C. garlic   D. cucumber
4. A. artichoke   B. lettuce   C. carrot   D. turnip
5. A. lettuce   B. tomato   C. cabbage   D. celery

Exercise 3 Match the key phrases with these pictures.(连线)

| match stick potatoes | straight potatoes | straw potatoes | crinkle potatoes |

Exercise 4 Write the sentences in the right order.(排序写句子)

1. cook, I, shall, how, spinach, the

2. I, shall, dice, onion, the

_____

3. slicing, green pepper, are, now, you, why

_____

4. is, the, most, cold dish, of, onion, "heart"

_____

5. for, chop, celery, the, dish, the, please

_____

Exercise 5 Fill in the blanks with given words.（选词填空）

| peel　　grind　　chop　　scoop... into　　balls　　batons　　slice |

1. _____ the potatoes firstly.

2. For the preparation work, _____ the garlic finely.

3. Please _____ the potato and zucchini _____ .

4. Would you cut some _____ onion for the soup?

5. ——Shall I chop the carrots?

　　——No, cut carrot _____ carefully.

6. ——What are you doing?

　　——I'm _____ the parsley.

Exercise 6 Translate the following sentences into Chinese or English.（中英文句子翻译）

1. Please wash the pumpkin, and then dice it.

_____

2. We need chopped onions to make the soup.

_____

3. 我该怎么挖土豆蔬菜球呢？

_____

4. 他正在捣蒜泥。

_____

5. 我现在剁欧芹末好吗？

_____

Unit 5
参考答案

Unit 6
Fruits & Nuts

You will be able to:

1. read and write the words of different fruits and nuts;
2. describe the ingredients of drinks and dishes;
3. talk about the process of making fruit or nut dishes.

Unit introduction 单元介绍

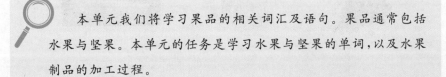

本单元我们将学习果品的相关词汇及语句。果品通常包括水果与坚果。本单元的任务是学习水果与坚果的单词,以及水果制品的加工过程。

扫码看课件

Thinking map（思维导图）

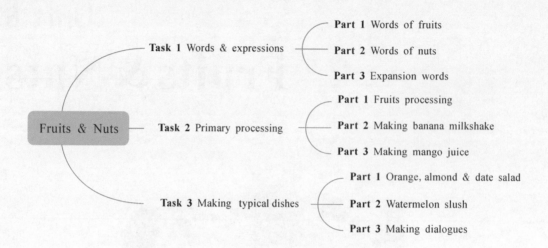

Warming up

Activity 1 Look and match. 看单词，连线。

cherry   pear   strawberry   peach

 Activity 2 Listen and tick. 听录音，勾出对话中要榨的果汁。

Task 1　Words & expressions 词汇

Part 1　Words of fruits. 水果类单词。

🎧 **Activity** Read and write. 读单词，看图写英文。

dragon fruit 火龙果　　fig 无花果　　kiwi fruit 猕猴桃　　passion fruit 百香果
pineapple 菠萝　　lychee 荔枝　　coconut 椰子　　star fruit 杨桃　　papaya 木瓜

_____　　_____　　_____

_____　　_____　　_____

_____　　_____　　_____

Part 2　Words of nuts. 坚果类单词。

🎧 **Activity 1** Listen and number. 听录音，标出正确顺序。

(　) peanut 花生　　(　) walnut 核桃　　(　) chestnut 栗子　　(　) hazelnut 榛子
(　) almond 杏仁　　(　) cashew 腰果　　(　) pine nut 松子　　(　) pistachio 开心果

Activity 2 Look and translate. 单词互译。

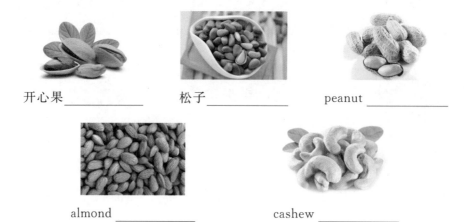

开心果_____   松子_____   peanut _____

almond _____   cashew _____

Activity 3 Look and write. 根据描述写出单词。

1. _____ A smooth red-brown nut, the color of inside is yellow.
2. _____ Small cream-colored seeds that grow on pine trees.
3. _____ A nut that you can eat, is shaped like a human brain.

Part 3   Expansion words. 拓展词汇。

Activity 1 Look and match. 看图连线。

葡萄   黑莓   李子   草莓   山莓   蓝莓   葡萄干   椰枣

blackberry   raisin   strawberry   date   raspberry   plum   grape   blueberry

Activity 2 Listen and write. 听录音写单词。

_____

_____

外教有声

Activity 3 Fill in the missing letters and make a self-evaluation. 补全单词并做自我评价。

1. p _ _ ch   2. r _ spberry   3. r _ _ sin   4. pl _ m   5. l _ chee
6. f _ g   7. p _ p _ ya   8. c _ sh _ w   9. _ lm _ nd   10. p _ st _ ch _ o

Assessment: If you can write 8-10 words, you are perfect.

                If you can write 4-7 words, you are good.

                If you can write 1-3 words, you will try again.

Task 2 Primary processing 初加工

Part 1 Fruit processing. 水果的加工。

Activity 1 Write down the tools for fruit processing. 写出水果加工所用工具的名称。

_____   _____   _____

_____   _____   _____

Activity 2 Look and choose. 看图并选择。

A. Cut the mango
B. Remove the core of the pineapple
C. Make orange juice
D. Slice the banana
E. Remove the ends of the strawberries
F. Get rid of the seeds of cherries

1. _____

2. _____

3. _____

4. _____

5. _____

6. _____

Activity 3 Listen and fill in the blanks. 听录音，完成对话。

A. slice the kiwi fruit
B. we'll make a fruit salad
C. a cup of yogurt in the salad
D. put all the fruits in a mixing bowl

Jack: What are we going to do today?

Chef: _____. Prepare one apple, one kiwi fruit, two bananas and some strawberries.

Jack: Oh, I see. And then?

Chef: _____, cut the bananas with the banana slicer. And then remove the base of the strawberries. At last, _____.

Jack: OK, It's that all?

Chef: Then put two teaspoons of honey and _____.

Jack: Mix them up?

Chef: That's right.

Part 2 Making banana milkshake. 香蕉奶昔的制作。

Activity 1 Look at the processing. 看香蕉奶昔的加工过程。

Activity 2 Listen and complete the dialogue. 听录音，参考 Activity 1 完成对话。

外教有声

| A. cut them | B. make banana milkshake |
| C. shall we do first | D. add some milk |

Chef: We are going to _____ today.

Jack: What _____ , chef?

Chef: Wash and peel two bananas, and then _____ , please.

Jack: Put them into the liquidizer?

Chef: Yes, then _____ .

Jack: After that?

Chef: Stir them for about 4 seconds. At last, add some vanilla ice-cream. It will be more beautiful and delicious.

Jack: Oh, I've got it. Sounds nice.

Part 3 Making mango juice. 杧果汁的制作。

Activity 1 Tick the tools for making mango juice. 勾出所需的工具。

Activity 2 Look and translate. 看图,翻译英文。

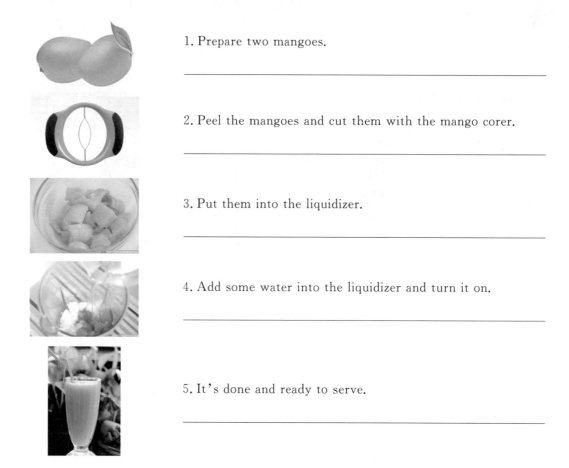

1. Prepare two mangoes.
   _____

2. Peel the mangoes and cut them with the mango corer.
   _____

3. Put them into the liquidizer.
   _____

4. Add some water into the liquidizer and turn it on.
   _____

5. It's done and ready to serve.
   _____

Activity 3 Make dialogues. 根据所给信息进行对话练习。

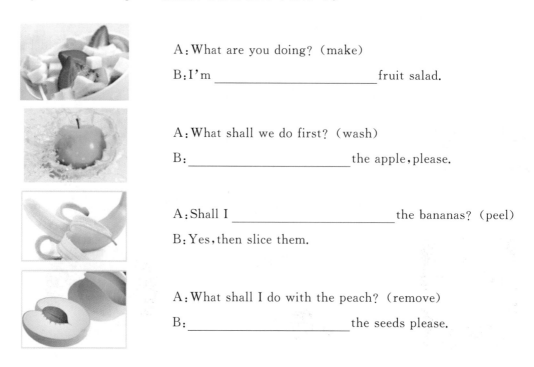

A: What are you doing? (make)
B: I'm _____ fruit salad.

A: What shall we do first? (wash)
B: _____ the apple, please.

A: Shall I _____ the bananas? (peel)
B: Yes, then slice them.

A: What shall I do with the peach? (remove)
B: _____ the seeds please.

Task 3　Making typical dishes
特色菜肴制作

Part 1　Orange, almond & date salad. 橘子杏仁椰枣沙拉。

> 橘子杏仁椰枣沙拉是一款具有养生功能的特色菜肴，橘子开胃理气，杏仁止咳润肺，椰枣健脾益胃。

Activity 1 Read the recipe. 读菜单。

外教有声

❶ **Ingredients**

orange
blood orange(血橙)
date
coconut oil
honey
cinnamon(肉桂)
mint
almond
salt

❷ **Methods**

1. Put the almonds in a small bowl and mix them into coconut oil, honey, cinnamon and salt.

2. Spread almonds in a single layer(一层) and bake about 10 minutes.

3. Cool the almonds before use.

4. Peel and slice the oranges and blood oranges, and then arrange(摆放) them on the plate.

5. Cutting the dates into pieces.

6. Scatter(撒入) dates, almonds, and mint on the top of the oranges.

Activity 2 Tick the ingredients. 勾出所需原料。

☐ orange　　　☐ coconut oil　　　☐ cinnamon　　　☐ almond
☐ salt　　　　☐ sugar　　　　　☐ honey　　　　 ☐ mint

Activity 3 Look and write. 看图写出制作流程。

　　A　　　　B　　　　C　　　　D　　　　E

A. _____
B. _____
C. _____
D. _____
E. _____

Part 2　Watermelon slush. 西瓜沙冰。

西瓜不仅水分多，营养也很丰富，含有蛋白质、糖、钙、磷、铁、钾、果糖、维生素A、维生素C、番茄红素，还含有人体必需的氨基酸。夏季吃西瓜有很好的解暑作用。

Activity 1 Read the dialogue. 读对话。

Chef: We'll make watermelon slush today, clean the blender first, then prepare one lemon and diced watermelon please.

Jack: OK, I've finished. How about lemon?

Chef: Peel the lemon and remove the seeds.

Jack: It's done, chef. And then?

Chef: Put them all in the blender, add 3 teaspoons of honey and blend with some ice.

Jack: For how long?

Chef: About 15 seconds.

Chef: At last, add some vanilla ice cream on the top of the slush.

Jack: Oh, sounds amazing.

Unit 6   Fruits & Nuts

Activity 2 Tool preparation. 工具准备。

砧板_____   水果刀_____   榨汁机_____

Activity 3 Ingredient preparation. 原料准备。

西瓜_____   柠檬_____   蜂蜜_____   冰块_____

Activity 4 Look and write the production process. 看图并写出制作流程。

1. Clean the blender.
2. Dice the watermelon.
3. Peel the lemon and remove the seeds.
4. Put them in the blender.
5. Add some honey.
6. Blend with ice cubes.

Part 3 Making dialogues. 对话练习。

A: What fruit would you like to eat?
B: I'd like to eat _____.

Necessary words and phrases（必需词汇）

cherry [ˈtʃeri] 樱桃
watermelon [ˈwɔːtəmelən] 西瓜
blood orange 血橙
melon [ˈmelən] 甜瓜
pineapple core [ˈpaɪnæpl kɔː(r)] 菠萝心儿
peach [piːtʃ] 桃子
grape [greɪp] 葡萄
pear [peə(r)] 梨
plum [plʌm] 李子
blackberry [ˈblækbəri] 黑莓
blueberry [ˈbluːbəri] 蓝莓
strawberry [ˈstrɔːbəri] 草莓
raspberry [ˈrɑːzbəri] 山莓
lychee [ˌlaɪˈtʃiː] 荔枝
papaya [pəˈpaɪə] 木瓜
kiwi fruit [ˈkiːwiːfruːt] 猕猴桃

passion [ˈpæʃn] fruit 百香果
fig [fɪg] 无花果
star fruit 杨桃
dragon fruit 火龙果
coconut [ˈkəʊkənʌt] 椰子
almond [ˈɑːmənd] 杏仁
peanut [ˈpiːnʌt] 花生
milkshake [ˈmɪlkʃeɪk] 奶昔
slush [slʌʃ] 沙冰
walnut [ˈwɔːlnʌt] 核桃
pistachio [pɪˈstæʃiəʊ] 开心果
chestnut [ˈtʃesnʌt] 栗子
cashew [ˈkæʃuː] 腰果
hazelnut [ˈheɪzlnʌt] 榛子
pine nut [paɪn nʌt] 松子
dates [deɪts] 椰枣

Expansion words and phrases（拓展词汇）

avocado [ˌævəˈkɑːdəʊ] 鳄梨，牛油果
haw [hɔː] 山楂

Unit 6　Fruits & Nuts

blackcurrant ['blækˌkʌrənt] 黑加仑
nectarine ['nektəri:n] 油桃
Chinese date 中国大枣
grapefruit ['greɪpfru:t] 西柚
loquat ['ləʊkwɒt] 枇杷

longan ['lɒŋgən] 龙眼
pomegranate ['pɒmɪgrænɪt] 石榴
apricot ['eɪprɪkɒt] 杏
lime [laɪm] 青柠

Notes（重点词汇、短语）

remove seeds	去籽
cut... into pieces	把……切成片
cool	冷却
spread	展开，铺开
arrange	摆放
scatter	撒入……
single layer	一层
be ready to serve	准备上桌
watermelon slush	西瓜沙冰
banana slicer	香蕉切片器
pineapple corer	菠萝取芯器
strawberry huller	草莓取蒂器
cherry pitter	樱桃去核器
mango corer	杧果去核器
citrus reamer	柑橘榨汁器

Culture life（文化生活）

各种果茶

蜂蜜红茶不仅闻起来清香怡人，更是一款受女士们青睐的养颜补血的最佳饮品。
配料：蜂蜜，红枣，生姜。

柠檬茶有生津止渴，化痰止咳，健脾，降糖消渴，巩固瘦身等效果。
配料：红茶，柠檬，蜂蜜。

水果茶可以舒缓情绪,并且含有大量维生素C,有美容养颜的效果。

配料:橘子,草莓,青柠,蜂蜜。

Practice 巩固练习

Exercise 1 Translate these words into Chinese or English.（中英单词互译）

1. peanut _____   2. 菠萝 _____

3. date _____   4. 木瓜 _____

5. chestnut _____   6. 荔枝 _____

7. walnut _____   8. 蓝莓 _____

9. plum _____   10. 杧果 _____

Exercise 2 Tick the different words.（找不同）

1. A. peach   B. apple   C. banana   D. peanut

2. A. chestnut   B. apricot   C. grape   D. peach

3. A. pineapple   B. coconut   C. star fruit   D. cherry

4. A. mango   B. papaya   C. apple   D. dragon fruit

Exercise 3 Find the answers.（找答案）

1. What are you doing?   A. Put them in the liquidizer.

2. What's the color of the grape?   B. Just so so, it's too sour.

3. Do you like plum?   C. I'm washing the fruits.

4. What shall we do with the melon?   D. It's purple.

5. What about the next?   E. Cut it into half and remove seeds.

Exercise 4 Try to summarize the fruits and nuts with your partners.（总结）

banana	orange	apple	grape	peach	coconut
pear	passion fruit	papaya	lychee	apricot	pine nut
plum	kiwi fruit	melon	walnut	chestnut	watermelon
cashew	strawberry	almond	raisin	peanut	cherry
star fruit	blueberry	fig	raspberry	blackberry	grapefruit

Tropical and subtropical fruits	Mainland fruits	Nuts
（热带与亚热带水果）	（大陆水果）	（坚果）
_____	_____	_____
_____	_____	_____
_____	_____	_____
_____	_____	_____
_____	_____	_____

Exercise 5 Complete the following sentences.（完成句子）

1. _____ a fruit salad.（做）

2. _____ the grapes.（浸泡）

3. _____ the pears.（去皮）

4. _____ the seeds from the melon.（去除）

5. Make _____.（香蕉奶昔）

6. Put the mango into the _____.（榨汁机）

7. The little girl likes _____ very much.（水果沙拉）

8. _____ the raisins on the salad.（撒入）

Exercise 6 Translate the following sentences into Chinese or English.（中英文句子翻译）

1. 今天我们要做猕猴桃沙冰。
_____

2. 请把椰枣洗净并浸泡半小时。
_____

3. 请用樱桃去核器去掉樱桃的核。
_____

4. We'll make strawberry shake today.
_____

5. Put the almonds in a small bowl and mix them into coconut oil, honey, cinnamon and salt.
_____

Unit 7
Meat

You will be able to:

1. know the words of different meat;
2. describe the meat preparation process;
3. talk about the process of making meat dishes.

Unit introduction 单元介绍

本单元我们将学习烹饪原料中的肉类产品。肉类产品通常包括家畜、家禽和野味等。本单元的任务是学习家畜、家禽类英文单词,用英文描述其初加工过程和以牛肉和鸡肉为主料的典型菜肴的制作过程。

扫码看课件

Thinking map（思维导图）

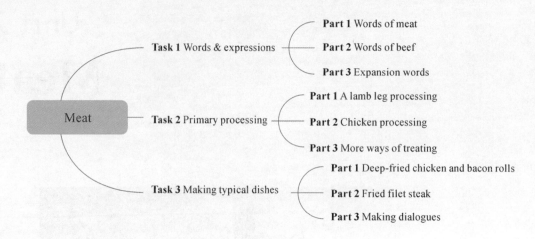

Warming up

Activity 1 Look and match. 看单词，连线。

chicken   beef   pork   lamb

外教有声

 Activity 2 Listen and tick. 听录音，勾出客人所点的食物。

Task 1　Words & expressions 词汇

Part 1　Words of meat. 肉类单词。

 Activity Read and write. 读单词，看图写英文。

turkey 火鸡　　pheasant 野鸡　　venison 鹿肉　　boar meat 野猪肉　　veal 小牛肉
veal kidney 牛腰子　　osso bucco 牛膝　　oxtail 牛尾肉　　foie gras 肥鹅肝

Part 2　Words of beef. 牛肉类单词。

 Activity 1 Listen and number. 听录音，标出正确顺序。

(　) chuck 牛肩胛　　(　) brisket 牛胸　　(　) rib 牛肋脊　　(　) shank 牛腱
(　) flank 牛腹肋　　(　) sirloin 牛西冷　　(　) tenderloin 牛里脊　　(　) round 牛后臀

Activity 2 Look and translate. 单词互译。

牛西冷_____

牛腱_____

tenderloin _____

brisket _____

chuck _____

Activity 3 Look and write. 根据描述写出单词。

1. _____ This part of beef is above the tenderloin.
2. _____ This part of beef is on the top of legs.
3. _____ This part of beef is surrounded by short loin, flank and round.

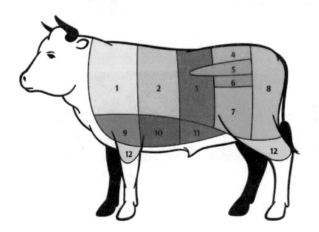

1—Chuck   7—Bottom sirloin
2—Rib   8—Round
3—Short loin   9—Brisket
4—Sirloin   10—Plate
5—Tenderloin   11—Flank
6—Top sirloin   12—Shank

Part 3 Expansion words. 拓展词汇。

Activity 1 Look and match. 看图连线。

煎   扒   炒   烤   炸   煮   烩   蒸

pan-fry   grill   sauté   roast   deep-fry   boil   stew   steam

Activity 2 Listen and write. 听录音写单词。

外教有声

_____

_____

Activity 3 Fill in the missing letters and make a self-evaluation. 补全单词并做自我评价。

1. v__l   2. oxt__l   3. ph____sant   4. turk____   5. fl__nk

6. ch__ck   7. sh__nk   8. s____loin   9. br__sk__t   10. r__b

Assessment: If you can write 8-10 words, you are perfect.

If you can write 4-7 words, you are good.

If you can write 1-3 words, you will try again.

Task 2 Primary processing 初加工

Part 1 A lamb leg processing. 羊腿的加工。

Activity 1 Write down the tools for removing bones from a lamb. 写出剔羊腿骨的工具名称。

_____   _____

Activity 2 Look and choose. 看图并选择。

A. Pull off the bone   B. Butterfly the lamb leg

C. Flatten the lamb   D. Cut the bone on the top

E. Take some fat out   F. Take out the bone inside

1. _____   2. _____   3. _____

4. _____   5. _____   6. _____

Activity 3 Listen and fill in the blanks. 听录音，完成对话。

A. cut the lamb leg open   B. flatten the lamb

| C. sharpening the bone knife | | D. mallet and bone knife |

Chef: Today you will learn how to remove bones from a lamb.

Jack: Cool. How do we call these tools in English?

Chef: They are called _____.

Jack: What can we do with the bone knife?

Chef: We can _____ and remove the bones with it.

Jack: Got it. How about next?

Chef: _____ with the mallet.

Jack: Is the steel used for _____?

Chef: Yes. Remember to keep the tools clean and dry.

Jack: Yes, I will.

Part 2 Chicken processing. 鸡肉的加工。

Activity 1 Look at the processes of cutting chicken. 看整只鸡的切割过程。

Activity 2 Listen and complete the dialogue. 听录音，参考 Activity 1 完成对话。

| A. remove the breast | | B. into different parts |

| C. remove the wings | | D. into two parts |

Jack: How do we cut the chicken _____, chef?

Chef: Remove one leg first, cut between ball and socket to separate it.

Jack: How do we cut the leg _____?

Chef: Cut the joint (关节) between thigh and drumstick.

Jack: And then _____?

Chef: Yes, cut along the back bone.

Jack: At last, we need to _____?

Chef: Right. Release a wing from breast, repeat the process once more on other side.

Part 3　More ways of treating. 更多加工方法。

Activity 1 Look and translate. 看图,翻译英文。

1. Soak the unfreezing foie gras in the milk for 8 hours.
_____

2. Put it in a plastic bag and heat it in water of 60 ℃.
_____

3. Heat it and then remove the grease in the bag.
_____

Activity 2 Make dialogues. 根据所给信息进行对话练习。

A: How can I cut the lamb? (butterfly)
B: _____ it with the knife.

A: What's this in English? (Osso Bucco)
B: It's called _____.

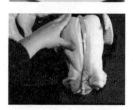

A: What _____ I _____ next? (do)
B: Cut along the back bone of the chicken.

A: Is _____ (grilled beef) your favorite dish?
B: Yes, it is.

Task 3　Making typical dishes
特色菜肴制作

Part 1　Deep-fried chicken and bacon rolls. 炸培根鸡肉卷。

此道菜为西餐常见热菜。用优选鸡胸肉或鸡腿肉裹上培根肉片烤制或煎炸。培根肉香扑鼻，金黄诱人，鸡腿肉鲜嫩多汁，咬上一口，美味难挡。

Activity 1 Read the recipe. 读菜单。

外教有声

❶ Ingredients

chicken breast
bacon
cucumber
mushroom
broccoli
red wine
thyme
mustard
lemon juice

❷ Methods

1. Slice the chicken breast and marinade with salt, red wine, ground black pepper and thyme.

2. Cut the vegetables into bars or pieces.

3. Roll the vegetables into the sliced chicken.

4. Cut the bacon into escalope and wrap in chicken rolls.

5. Heat oil in the frying pan and then pan-fry the rolls.

Activity 2 Tick the ingredients. 勾出所需原料。

☐ bacon ☐ pepper powder ☐ lemon juice ☐ mustard
☐ broccoli ☐ cucumber ☐ chicken breast ☐ thyme

Activity 3 Look and write. 看图写出制作流程。

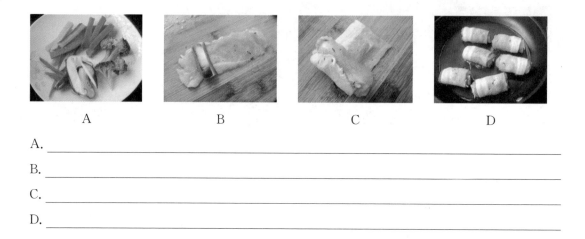

A　　　　　　　B　　　　　　　C　　　　　　　D

A. _____
B. _____
C. _____
D. _____

Part 2　Fried filet steak. 煎菲力牛排。

> 菲力牛排用的是牛里脊最嫩的部分，肉质鲜美多汁。菲力牛排可以煎成三分熟（rare），四分熟（medium rare），五分熟（medium），七分熟（medium well）和十分熟（well done）。

Activity 1 Read the dialogue. 读对话。

Chef: We'll learn to make fried filet steak.

Jack: Sounds great. What do we need to prepare?

Chef: First of all, we need to prepare cooking utensils such as pan, cook knife, tenderizer, barbecue clip and plate.

Jack: Are these ingredients for making fried filet steak?

Chef: Yes, do you know how to call them in English?

Jack: Filet beef, red wine, brandy, butter, olive oil, ground black pepper, salt, garlic powder, minced garlic, onions and carrots.

Chef: Well done, Jack, let's start.

Activity 2 Tool preparation. 工具准备。

盆_____

厨刀_____

刷子_____

盘子_____

烤肉夹_____

横纹锅_____

Activity 3 Ingredient preparation. 原料准备。

菲力牛肉_____

胡萝卜_____

洋葱_____

黄油_____

大蒜_____

盐_____

橄榄油_____

黑胡椒粉_____

红酒_____

Activity 4 Look and write the production process. 看图写出制作流程。

1. Prepare filet beef and a grill pan.
2. Chop the garlic into mince and prepare some garlic powder.
3. Melt the butter in a pot and mix the garlic powder with it.
4. Marinate the beef with salt, onions, carrots and red wine, pan-fry with olive oil and spread some black pepper powder on it at last.
5. Season the cooked steak with minced garlic.
6. Transfer to warmed plates and rest for 2 to 3 minutes before serving.

_____   _____   _____

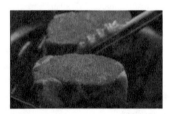  

_____   _____   _____

Part 3 Making dialogues. 对话练习。

A: What would you like to have, dad?

B: I'd like _____.

A: Do you know how to say it in Chinese?

B: _____.

| Grilled New Zealand lamb chops 扒新西兰羊排 | Curry chicken 咖喱鸡 | Fried spire ribs 炸猪排 |

Unit 7　Meat

Necessary words and phrases（必需词汇）

filet steak 菲力牛排
rib eye steak 肉眼牛排
filet [ˈfɪlɪt] 菲力（牛里脊）
chuck [tʃʌk] 牛肩胛
beef rib [rɪb] 牛肋脊
flank [flæŋk] 牛腹肋
turkey [ˈtɜːki] 火鸡
pheasant [ˈfeznt] 野鸡
boar [bɔː(r)] 野猪

T-bone steak T 骨牛排
sirloin steak 西冷牛排
tenderloin [ˈtendərlɔɪn] 牛里脊
brisket [ˈbrɪskɪt] 牛胸
shank [ʃæŋk] 牛腱
round [raʊnd] 牛后臀
venison [ˈvenɪsn] 鹿肉
veal [viːl] 小牛肉
veal kidney [ˈkɪdni] 小牛腰子

外教有声

Expansion words and phrases（拓展词汇）

foie gras 鹅肝
oxtail [ˈɒksteɪl] 牛尾肉
quail [kweɪl] 鹌鹑
chicken thigh [ˈtʃɪkɪn θaɪ] 鸡大腿
chicken breast [brest] 鸡胸
lamb leg [læm leg] 羊腿肉
lamb rack [ræk] 羊架肉
dice [daɪs] 切丁
shred [ʃred] 切丝

osso bucco 牛膝
pigeon [ˈpɪdʒɪn] 鸽子
guinea fowl [ˈɡɪni faʊl] 珍珠鸡
drumstick [ˈdrʌmstɪk] 琵琶腿
chicken wing [wɪŋ] 鸡翅
lamb chop [tʃɒp] 羊排肉
lamb loin [lɔɪn] 羊腰脊
slice [slaɪs] 切片
chop [tʃɒp] 剁碎

外教有声

Notes（重点词汇、短语）

Task 1

short loin 前腰脊
bottom sirloin 下后腰脊

top sirloin 上后腰脊
plate 胸腹

Task 2

butterfly the lamb　　　　　　　　　　把羊腿蝶翅形切开
sharpening the bone knife　　　　　　把厨刀磨锋利
remove the bones from the lamb leg　　从羊腿中剔除骨头
mallet and bone knife　　　　　　　　木槌和剔骨刀
keep... clean and dry　　　　　　　　保持……清洁干燥
between ball and socket　　　　　　　鸡腿球状肉和关节窝之间

separate... from	把……分离
release... from	把……卸掉
remove the grease	去掉油脂

Task 3

mix... with	把……混合
marinate... with	用……腌渍
season... with	用……调味

Culture life（文化生活）

西餐牛排的分类知识

牛排也称牛扒，是西餐中最常见的食物之一。欧洲中世纪时，猪肉及羊肉是平民百姓的食用肉，牛肉则是王公贵族们的高级肉品，搭配上胡椒及香辛料一起烹调，并在特殊场合中供应，以彰显主人的尊贵身份。牛排的种类非常多，常见的有以下四种。

【菲力牛排】菲力又称为牛里脊或嫩牛柳，是牛脊上最嫩的肉，受到爱吃瘦肉的人的青睐。

【肉眼牛排】肉眼是指牛肋上的肉，瘦肉和肥肉兼而有之，由于含一定肥膘，这种肉煎烤味道比较香。烹饪时最好不要煎得过熟，三分熟最好。

【西冷牛排】西冷牛排又称为沙朗牛排，西冷是牛外脊上的肉，含一定肥油，在肉的外沿带一圈呈白色的肉筋，总体口感韧度强、肉质硬、有嚼头，适合年轻人和牙口好的人吃。切肉时连筋带肉一起切，另外不要煎得过熟。

【T骨牛排】T骨牛排是牛背上的脊骨肉，呈T字形，两侧一边是菲力，另一边是西冷，人们既可以尝到菲力牛排的鲜嫩，又可以感受到西冷牛排的芳香，一举两得。此种牛排在美式餐厅更常见，由于法餐讲究制作精致，对于量较大而肉质较粗糙的T骨牛排较少采用。

T骨牛排

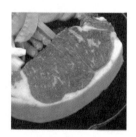
西冷牛排

肉眼牛排

菲力牛排

Practice 巩固练习

Exercise 1 Translate these words into Chinese or English.（中英单词互译）

1. 火鸡_____
2. foie gras _____
3. 鹿肉_____
4. osso bucco _____

Unit 7　Meat

5. 牛腱_____　　　　　　6. pigeon _____

7. 排骨_____　　　　　　8. pheasant _____

9. 小牛肉_____　　　　　10. brisket _____

Exercise 2 Tick the different words.（找不同）

1. A. round　　　　　　B. chuck　　　　　　C. flank　　　　　　D. boar

2. A. chicken thigh　　B. chicken breast　　C. chicken wing　　D. chicken

3. A. short rib　　　　B. chuck　　　　　　C. sirloin steak　　D. tenderloin

4. A. chop　　　　　　B. slice　　　　　　 C. shred　　　　　　D. lamb

5. A. oxtail　　　　　　B. guinea fowl　　　C. pigeon　　　　　D. quail

Exercise 3 Choose the right phrases for the pictures.（用正确词组填空）

A. cut the lamb open　　　B. flatten the lamb　　　C. butterfly the lamb

D. take the fat out　　　　E. remove the bone

_____　　　　_____　　　　_____

_____　　　　_____

Exercise 4 Write the sentences in the right order.（排序写句子）

1. how, you, from a lamb, will, to remove bones, learn

_____

2. are called, the tools, steel and knife

_____

3. clean and dry, remember, the knife, to keep, please

_____

4. to separate the chicken leg, cut, between ball and socket

_____

5. from the chicken breast, release, at last, the chicken wings

Exercise 5 Match the English phrases with the Chinese one.（连线翻译西餐菜名）

Deep-fried chicken and bacon rolls　　　　扒新西兰羊排
Fried sirloin steak　　　　　　　　　　　咖喱鸡
Grilled New Zealand lamb chops　　　　　炸猪排
Fried spare ribs　　　　　　　　　　　　炸培根鸡肉卷
Curry chicken　　　　　　　　　　　　　煎西冷牛排

Exercise 6 Answer the following questions.（回答问题）

1. What is the steel used for?

2. What can we use to remove the bones?

3. Do we usually separate the chicken with the cook knife?

4. What kind of beef do we use to make fried filet steak?

5. How do you call "deep-fried chicken and bacon rolls" in Chinese?

Exercise 7 Translate the following sentences into Chinese or English.（中英文句子翻译）

1. Soak unfreezing foie gras in the milk for 8 hours.

2. Heat it and then remove the grease in the bag.

3. 我们今天学习制作煎菲力牛排。

4. 这些调料是用来做什么的？

5. 把煎熟的牛排用大蒜末调味。

Unit 7 参考答案

Unit 8
Seafood

You will be able to:

1. know the words of different seafood;
2. describe the seafood preparation process;
3. talk about the process of making seafood dishes.

Unit introduction 单元介绍

本单元我们将学习烹饪原料中的海产品。海产品通常包括鱼类、虾类和贝类等。本单元的任务是学习鱼虾贝类单词、初加工过程和以鱼、虾为主料的典型菜肴的制作过程。

扫码看课件

Thinking map（思维导图）

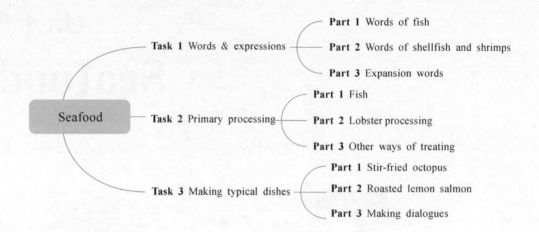

Warming up

Activity 1 Look and match. 看单词，连线。

tuna   crab   shrimp   salmon

外教有声

Activity 2 Listen and tick. 听录音，勾出小明晚餐吃的食物。

Task 1　Words & expressions 词汇

Part 1　Words of fish. 鱼类单词。

 Activity Read and write. 读单词，看图写英文。

cuttlefish 墨鱼	sardine 沙丁鱼	sole 比目鱼	octopus 章鱼
codfish 鳕鱼	plaice 鲽鱼	red trout 虹鳟鱼	sea bass 海鲈鱼

外教有声

_____

_____

_____

_____

_____

_____

_____

_____

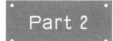

Part 2　Words of shellfish and shrimps. 虾蟹类单词。

Activity 1 Listen and number. 听录音，标出正确顺序。

(　) lobster 龙虾　　(　) prawn 明虾　　(　) shrimp 小虾　　(　) shrimp meat 虾仁
(　) oyster 牡蛎　　(　) scallop 扇贝　　(　) abalone 鲍鱼　　(　) snail 蜗牛
(　) mussel 蚌，贻贝　(　) king crab 帝王蟹

Activity 2 Look and translate. 单词互译。

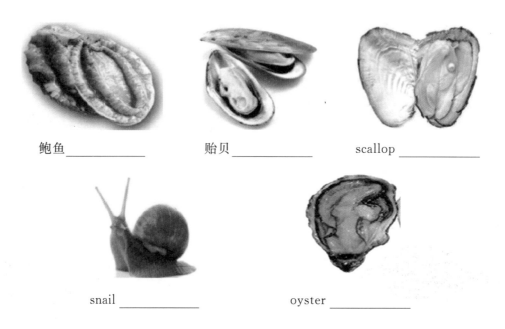

鲍鱼_____　　　贻贝_____　　　scallop_____

snail_____　　　oyster_____

Activity 3 Look and write. 根据描述写出单词。

1. _____ It has claws only on two pairs of legs which are shorter than prawn.
2. _____ It has claws on three pairs of their legs which are larger than shrimps.
3. _____ It has long bodies with muscular（强健的）tails and live on the sea floor.
4. _____ It is the meat of shrimp and often cooked in cold and hot dishes.

Part 3　Expansion words. 拓展词汇。

Activity 1 Look and match. 看图连线。

红鲱鱼　鳜鱼　鲷鱼　安康鱼　鱼子酱　鱿鱼　蛤　海螺

snapper　caviar　squid　clam　anglerfish　sea snail　red herring　mandarin fish

 Activity 2 Listen and write. 听录音写单词。

_____
_____

外教有声

Activity 3 Fill in the missing letters and make a self-evaluation. 补全单词并做自我评价。

1. s __ __ dine　2. c __ df __ sh　3. red tr __ __ t　4. c __ vi __ r　5. sea sn __ il
6. lobst __ __　7. oyst __ __　8. __ b __ lone　9. m __ ss __ l　10. __ ngl __ rfish

Assessment: If you can write 8-10 words, you are perfect.

If you can write 4-7 words, you are good.

If you can write 1-3 words, you will try again.

Task 2　Primary processing 初加工

Part 1　Fish processing. 鱼的加工。

Activity 1 Write down the tools for fish processing. 写出鱼类加工所用的工具名称。

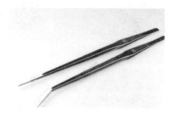

_____　　_____　　_____

Activity 2 Look and choose. 看图并选择。

A. Clean the fish　　　　B. Gut the gills

C. Take out the gills　　　D. Cut open the fish

E. Scale the fish

1. _____　　2. _____　　3. _____

4. _____　　5. _____

Unit 8　Seafood

Activity 3 Listen and fill in the blanks. 听录音，完成对话。

A. scale the fish with the scale

B. fish scale and fish scissors

C. use the fish scissors to gut

D. get rid of the fish line

Chef: Today I'll teach you how to prepare the sea bass.

Jack: Wonderful. What do we call these fish tools in English, chef?

Chef: They are _____.

Jack: How shall we do with them?

Chef: You can _____ and take out the gills with the scissors.

Jack: I've done it. How about next?

Chef: _____ the sea bass.

Jack: Oh, these fish tools are very useful.

Chef: Yes. Don't forget to _____ and clean it.

Jack: I see.

Part 2　Lobster processing. 虾产品加工。

Activity 1 Look at the processing. 看龙虾的加工过程。

Activity 2 Listen and complete the dialogue. 听录音，参考 **Activity 1** 完成对话。

A. with your thumbs

B. remove the meat

C. take hold of the tail

D. cut down the center

Jack: What shall we do with the lobster before cooking, chef?

Chef: We should _____ and twist it.

Jack: OK. it's so easy. And then?

Chef: _____ with the sharp kitchen scissors.

Jack: Shall we use the lobster crackers next?

Chef: Yes, we can _____ from them.

Jack: Could I use a little hammer instead of lobster cracker?

Chef: No problem.

Part 3　More ways of treating. 更多加工方法。

Activity 1 Look and translate. 看图，翻译英文。

1. Slice across the head to the backbone, separate the fillet from the head.

_____

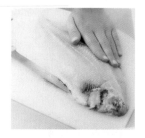

2. Cut along one side of the fish.

_____

3. Cutting to the other side, until the whole fillet is released.

_____

4. Lift off the fillet in a single piece and then turn the fish over to repeat the process.

_____

Unit 8 Seafood

Activity 2 Make dialogues. 根据所给信息进行对话练习。

A: What should I do for the next step? (gill)
B: You should _____.

A: What're you doing now? (cut... steak)
B: I'm _____.

A: What shall I _____ the fish? (fillet)
B: You can use the knife.

A: Do you like eating _____? (pan-fried fish)
B: Yes, so much.

Task 3 Making typical dishes
特色菜肴制作

Part 1 Stir-fried octopus. 深炸章鱼。

章鱼是高蛋白低脂肪的食材,含有丰富的蛋白质、脂肪、钙、铁、锌等营养成分。章鱼富含牛磺酸,能抗疲劳、降血压及软化血管。

Activity 1 Read the recipe. 读菜单。

❶ Ingredients

octopus
flour
chili powder
garlic
sugar
light soy sauce
ginger juice
ground white pepper
cooking oil
onion
carrot
zucchini
leek
red chili sesame oil
white sesame seeds

❷ Methods

1. Pour flour onto octopus, and then use hands to rub octopus to clean it.
2. Cut octopus into 5 cm (2 inches) pieces.
3. Heat oil and fry chopped garlic and onion until fragrant (香味).
4. Stir-fry octopus briefly over high heat. Don't overcook, or octopus will be tough.
5. Garnish with leek and sesame seeds before serving.

Activity 2 Tick the ingredients. 勾出所需原料。

☐ octopus ☐ cooking oil ☐ white wine ☐ flour
☐ salt ☐ sugar ☐ chili powder ☐ garlic

Activity 3 Look and write. 看图写出制作流程。

A

B

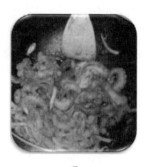
C

A. _____
B. _____
C. _____

Part 2　　Roasted lemon salmon. 柠檬烤三文鱼。

> 柠檬具有促消化、清肠排毒、美白肌肤、防癌抗癌、提高免疫力的作用,配上三文鱼,好吃的味道会带给你令人兴奋的舌尖触感。

Activity 1 Read the dialogue. 读对话。

Jack: I will make roasted lemon salmon. What do I need?

Chef: You should prepare the tools first. What do you think of that?

Jack: Cutting board, kitchen knife, oven, frying pan and plate are necessary.

Chef: Wonderful. Don't forget the tinfoil.

Jack: Oh, I almost forget it. How about the ingredients?

Chef: Salmon, lemon, cherry tomatoes, onion, sweet beans, rosemary, olive oil, salt and black pepper powder.

Jack: I've finished them, chef.

Chef: Now let's make the dish.

外教有声

Activity 2 Tool preparation. 工具准备。

砧板_____

锡纸_____

烤箱_____

菜盘_____

分刀_____

炒锅_____

Activity 3 Ingredients preparation. 原料准备。

三文鱼_____

柠檬_____

圣女果_____

洋葱_____

甜豆_____

迷迭香_____

盐_____

黑胡椒粉_____

橄榄油_____

Activity 4 Look and write the production process. 看图写出制作流程。

1. Preheat your oven to 190 ℃.

2. Slice lemon and garlic, chop onion, sweet beans to bar.

3. Sprinkle some rosemary, salt and black pepper powder on salmon.

4. Put sliced lemon, salmon on the tinfoil and wrap them together.

5. Roast them in the oven for 10-15 minutes, and then take them out.

6. Fry them and add sweet beans and cherry tomatoes, sprinkle little.

Part 3　Making dialogues. 对话练习。

A: What is your characteristics of seafood, waiter?
B: It is _____.
A: How do we call it in Chinese?
B: It is called _____.

Stewed sea bass in white sauce　　Smoked salmon　　Pan-fried cod
白汁海鲈鱼　　　　　　　　　　　烟熏三文鱼　　　　香煎鳕鱼

Necessary words and phrases（必需词汇）

salmon ['sæmən] 三文鱼　　　　　　lobster ['lɒbstə(r)] 龙虾

cuttlefish ['kʌtlfɪʃ] 墨鱼　　　　　　shrimp [ʃrɪmp] （小）虾

外教有声

sardine [sɑːˈdiːn] 沙丁鱼
sole [səʊl] 比目鱼
octopus [ˈɒktəpəs] 章鱼
codfish [ˈkɒdfɪʃ] 鳕鱼
plaice [pleɪs] 鲽鱼
red trout [traʊt] 虹鳟鱼
sea bass [beɪs] 海鲈鱼

prawn [prɔːn] 明虾
king crab 帝王蟹
oyster [ˈɔɪstə(r)] 牡蛎
scallop [ˈskɒləp] 扇贝
abalone [ˌæbəˈləʊni] 鲍鱼
mussel [ˈmʌsl] 贻贝，蚌
clam [klæm] 河蚌，蛤

外教有声

Expansion words and phrases（拓展词汇）

pomfret [ˈpɒmfrɪt] 平鱼
pike [paɪk] 梭子鱼
eel [iːl] 鳗
herring [ˈherɪŋ] 鲱鱼
yellow croaker [ˈkrəʊkə(r)] 黄花鱼
turbot [ˈtɜːbət] 多宝鱼
mackerel [ˈmækrəl] 马鲛鱼

sea snail [sneɪl] 海螺
carp [kɑːp] 鲤鱼
grass carp 草鱼
sturgeon [ˈstɜːdʒən] 鲟鱼
golden carp 鲫鱼
tilapia [tɪˈleɪpɪə] 罗非鱼
grouper [ˈgruːpə(r)] 石斑鱼

Notes（重点词汇、短语）

Task 1

fish scale	刮鳞器
fish scissors	鱼剪刀
fish tweezers	鱼镊子
take out the gills	取出鱼鳃
get rid of	拿出，放弃
forget to do	忘记做……
lobster cracker	蟹钳
little hammer	小锤子

Task 2

take hold of	拧下来，拧掉
across the head to the backbone	从头到背部骨头
cut along one side	沿着一边切
lift off	举起，拿起
fillet the fish	切鱼片
cut... steak	把……切成鱼排

Task 3

sprinkle some rosemary　　　　　　　　撒一些迷迭香
characteristics of seafood　　　　　　　特色海鲜

Culture life（文化生活）

三文鱼的基本知识

【三文鱼的种类】

帝王三文鱼(chinook salmon)、虹鳟(rainbow trout)、红三文鱼(sockeye salmon)、粉三文鱼(pink salmon)、马苏三文鱼(masu salmon)、阿拉斯加三文鱼(Alaskan salmon)和挪威三文鱼(Norwegian salmon)。

【三文鱼的挑选】

新鲜的三文鱼摸上去有弹性，按下去会慢慢恢复，不新鲜的三文鱼则相反。如果离鱼肉足够近，那么可以闻一闻，除了一点点咸咸的海水气息不应该有其他过于强烈的味道。含水量是三文鱼新鲜度和保存环境的重要指标。如果在鱼肉的边缘或者表面看到棕色的点，那么一定不能买。

【三文鱼的烹饪技法】

三文鱼的吃法多种多样，最常见的就是做成刺身，而有经验的师傅对于处理这种橙白相间、丰腴柔美的鱼类，无论是将其制作成寿司刺身生食还是煎烤成熟食，都能让人馋涎欲滴。

【三文鱼菜肴】

低卡路里的柠檬三文鱼

照烧三文鱼配西葫芦

辣烤三文鱼搭配柠檬蒜泥蛋黄酱

Practice 巩固练习

Exercise 1 Translate these words into Chinese or English.（中英单词互译）

1. sea bass _____
2. 牡蛎 _____
3. sardine _____
4. 贻贝 _____
5. mandarin fish _____
6. 鲍鱼 _____
7. snapper _____
8. 章鱼 _____
9. yellow croaker _____
10. 三文鱼 _____

Exercise 2 Tick the different words.（找不同）

1. A. fish scissors   B. fish scale   C. pomfret   D. fish forceps
2. A. shrimp   B. mussel   C. prawn   D. lobster
3. A. take out   B. cut open   C. scale   D. scallop
4. A. delicious   B. terrible   C. well   D. wonderful
5. A. vinegar   B. black pepper   C. soy sauce   D. crab

Exercise 3 Match the key phrases with these pictures.（连线）

clean the fish   take out the gills   gut the fish

scale the fish   cut open the fish

Exercise 4 Write the sentences in the right order.（排序写句子）

1. on, the, fish, fry, for, 1 minute, both sides, about

_____

2. scale, with, the fish scale, the fish

_____

3. well, wash, put into, the, a basin, shrimps, and

_____

4. with kitchen scissors, from the head, cut the backbone, to tail ends

_____

5. what, shellfish, you, to eat, would, of, kind, like

_____

Exercise 5 Translate western recipe.（翻译西餐菜单）

Grilled shrimps with cheese   Grilled red snapper fillet

Roasted salmon steak with tagliatelle and saffron sauce

Grilled fish fillet in lemon butter sauce

1. 煎红鲷鱼排_____

2. 芝士焗大虾_____

3. 黄油柠檬汁扒鱼柳_____

4. 烤三文鱼排、意大利宽面和藏红花汁_____

Exercise 6 Find the answers.（找答案）

1. What do we call these fish tools?   A. Remove the guts using your hands.

2. Shall I preheat the oven?   B. We'll cook pan-fried salmon.

3. What do I remove the guts?   C. Yes, preheat the oven to 190 ℃.

4. What shall we do today?   D. For about 15 minutes.

5. How long will you steam the dish?   E. They are fish scale and fish scissors.

Exercise 7 Translate the following sentences into Chinese or English.（中英文句子翻译）

1. Garnish with leek and sesame seeds before serving.

_____

2. Remove the bass from the wok and serve on a plate.

_____

3. 不同的工具有不同的用途。

_____

4. 请用这把特别的剪刀剪开鱼腹。

_____

5. 麦克和师傅正在做澳洲大龙虾。

_____

Unit 8
参考答案

Unit 9
Sauce & Soup

You will be able to:

1. get familiar with different kinds of sauce and soup;
2. know the basic sentences about making sauce and soup;
3. talk about the process of making sauce and soup.

Unit introduction 单元介绍

本单元我们将介绍西餐中基础汤汁的制作,其中少司是指经过厨师专门制作的菜点调味汁。本单元的任务是学习与汤汁相关的英文单词和用英文描述少司和汤类典型菜肴的制作过程。

扫码看课件

Thinking map (思维导图)

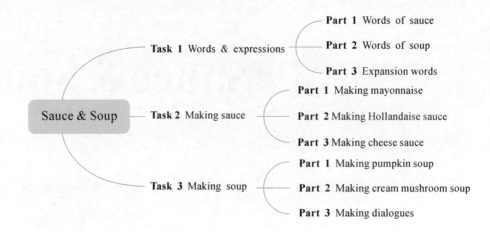

Warming up

Activity 1 Look and match. 看单词，连线。

| tomato sauce | mayonnaise | curry sauce | vinaigrette |

外教有声

Activity 2 Listen and tick. 听录音，勾出汤姆正在喝的汤。

Task 1　Words & expressions 词汇

Part 1　Words of sauce. 少司单词。

 Activity Read and write. 读单词, 看图写英文。

外教有声

Hollandaise sauce 荷兰少司　　Thousand Island sauce 千岛汁　　chocolate sauce 巧克力少司
cranberry sauce 蔓越莓少司　　caviar sauce 鱼子少司　　tartar sauce 鞑靼少司

Part 2　Words of soup. 西餐汤的名称。

 Activity 1 Listen and number. 听录音, 标出正确顺序。

外教有声

(　) cream asparagus soup 奶油芦笋汤　　　(　) cream pumpkin soup 奶油南瓜汤

(　) beef consommé 牛肉清汤　　　　　　(　) potato cheese soup 土豆芝士浓汤

(　) green pea puree soup 青豆蓉汤　　　　(　) tomato soup 番茄汤

Activity 2 Look and translate. 单词互译。

牛肉清汤_____

奶油芦笋汤_____

奶油南瓜汤_____

青豆蓉汤_____

Activity 3 Read and decide true(T) or false(F). 根据提示判断正误。

The word sauce is a French word that means a relish（调味汁）to make our food more appetizing. Sauces are liquid（液体）or semi-liquid foods devised to make other foods look, smell and taste better.

1. _____ The word sauce is a German word.
2. _____ Sauce means relish.
3. _____ Sauces are solid foods to make other foods look and taste better.

Part 3   Expansion words. 拓展词汇。

Activity 1 Look and match. 看图连线。

红酒少司   橘子少司   意大利蔬菜浓汤   蘑菇汤

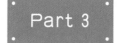

red wine sauce   orange sauce   Minestrone soup   mushroom soup

Unit 9   Sauce & Soup

 Activity 2 Listen and write. 听录音写单词。

_____

_____

Activity 3 Fill in the missing letters and make a self-evaluation. 补全单词并做自我评价。

1. s __ __ ce   2. c __ rry   3. __iqu __ d   4. may ___ nn __ __ se
5. r __ l __ sh   6. l __ bster   7. v __ n __ gar   8. c __ ns __ mmé

Assessment: If you can write 8 words, you are perfect.

  If you can write 4-7 words, you are good.

  If you can write 1-3 words, you will try again.

Task 2   Sauce making 少司的制作

Part 1   Making mayonnaise. 制作蛋黄酱。

Activity 1 Write down the tools for making mayonnaise. 写出制作蛋黄酱所用的工具名称。

_____   _____

Activity 2 Look and choose. 看图并选择。

A. Put the egg yolks into a bowl

B. Keep whisking until the mixture is thickened, and then it's ready to serve

C. Add 300 mL oil and continue to whisk

D. Stop whisking when the mixture is frothy(多泡的)

E. Mix the ingredients with a whisk

F. Put lemon juice, mustard, vinegar, salt and pepper into the mixing bowl

1. _____

2. _____

3. _____

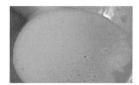
4. _____

5. _____

6. _____

Activity 3 Listen and fill in the blanks. 听录音，完成对话。

A. add the oil slowly

B. stop when the mixture is frothy

C. put the egg yolks into a bowl

D. mix the ingredients with a whisk

Jack: I have prepared 2 egg yolks, lemon juice, mustard, vinegar, salt and pepper. Can you show me how to make mayonnaise?

Chef: Sure. First, _____. And add the lemon juice, mustard, vinegar, salt and pepper. Then _____.

Jack: When shall I stop whisking?

Chef: _____ and you should _____ now. Then keep whisking.

Jack: How do I know the mayonnaise is ready?

Chef: Whisk until the sauce is thickened and it is ready to serve.

Jack: Oh, I see. Thank you.

Part 2 Making Hollandaise sauce. 制作荷兰少司。

Activity 1 Look. 了解制作荷兰少司的原料。

Unit 9 Sauce & Soup

Activity 2 Listen and complete the dialogue. 听录音，参考 **Activity 1** 完成对话。

A. the egg yolks

B. take the pan off

C. to your desired taste

D. first cook the butter

Chef: Have you got all the ingredients ready?

Jack: Yes. What shall I do with them?

Chef: _____ with low heat. Then remove the foam from the surface. Strain the butter into a jug.

Jack: What next?

Chef: Place _____ and 3 tablespoons of white wine into the mixing bowl and whisk them quickly. _____ the heat and start to add the lemon juice and clarified butter slowly.

Jack: The sauce becomes thick. What shall I do next?

Chef: Add salt and pepper _____ , the Hollandaise sauce is completed.

Jack: Wow, it's amazing.

Part 3   Making cheese sauce. 制作芝士少司。

Activity 1 Tick the tools for making cheese sauce. 勾出制作芝士少司的工具。

Activity 2 Look and translate. 看图，翻译英文。

1. Cook the butter.
   _____

2. Add flour, nutmeg, salt and pepper and whisk.
   _____

3. Keep whisking and add some cheese.

4. Serve the broccoli with cheese sauce.

Task 3　Making soup 汤菜制作

Part 1　Making pumpkin soup. 制作南瓜汤。

南瓜的营养极其丰富,南瓜含有淀粉、蛋白质、胡萝卜素、维生素B、维生素C、钙、磷等成分。南瓜所含成分能促进胆汁分泌,加强胃肠蠕动,帮助食物消化。

Activity 1 Read the recipe. 读菜单。

❶ **Ingredients**

chicken stock
salt
pumpkin puree
fresh parsley
onion
garlic
thyme leaves
ground pepper

❷ **Methods**

1. Chop the onions and parsley.

2. Heat stock. Put the onion, thyme leaves, garlic and ground pepper into the stock.

3. Puree the soup with a blender(搅拌器).

4. Reduce heat to low and simmer for another 30 minutes, uncovered, season with salt.

5. Pour into soup bowls and garnish(装饰、点缀) with fresh parsley.

Activity 2 Tick the ingredients. 勾出所需原料。

☐ onion ☐ pumpkin ☐ parsley ☐ garlic
☐ salt ☐ sugar ☐ ground pepper ☐ stock

Activity 3 Look and write. 看图写出制作流程。

A B C

A. _____
B. _____
C. _____

Part 2 Making cream mushroom soup. 制作奶油蘑菇汤。

奶油蘑菇汤是法国菜谱之一，以蘑菇为制作主料，奶油蘑菇汤的烹饪技巧以白烧为主，口味属于奶汤咸鲜。

 Activity 1 Read the dialogue. 读对话。

Jack: Chef, could you tell me how to make cream mushroom soup?

Chef: Sure. First, slice the mushroom, onion and garlic. Then place the pan over a low heat and melt the butter.

Jack: It's done. Shall I put the vegetable slices into the pan?

Chef: Yes. When the vegetables are soft, add chicken stock and mix it with a wooden spoon. Then add the batter.

Jack: I see. What next?

外教有声

Chef: Then cook for 25 minutes, add cream, season with salt and pepper.

Jack: Thank you. I can't wait to taste it.

Activity 2 Tool preparation. 工具准备。

木勺_____   分刀_____   汤锅_____

Activity 3 Ingredient preparation. 原料准备。

蘑菇_____   洋葱_____   大蒜_____

面粉_____   牛奶_____   鸡汤_____

盐_____   黑胡椒粉_____   黄油_____

Activity 4 Look and write the production process. 看图写出制作流程。

1. Melt butter in a wide, shallow pot.
2. Mix milk and flour, and then pour the mixture into the soup.

3. Season with black pepper powder and salt, and then serve while warm.

4. Put the mushrooms, garlic and onion into the pot.

5. Slice the mushrooms, onion and garlic.

6. Pour in the chicken stock and boil.

_____   _____   _____

_____   _____   _____

Part 3 Making dialogues. 对话练习。

A: Do you know what sauce it is?

B: Yes, it is _____.

A: What is it for?

B: It is for _____.

rosemary sauce　　　　　lemon sauce　　　　　Bearnaise sauce
迷迭香少司　　　　　　柠檬少司　　　　　　班尼士少司

Necessary words and phrases（必需词汇）

sauce [sɔːs] 少司　　　　　　　　puree [ˈpjʊəreɪ] 蓉汤

mayonnaise [ˌmeɪə'neɪz] 蛋黄酱
curry ['kʌri] 咖喱
simmer ['sɪmə(r)] 炖
consommé [kən'sɒmeɪ] 清汤

frothy ['frɒθi] 多泡的
garnish ['ɡɑːnɪʃ] 装饰
rosemary ['rəʊzməri] 迷迭香

外教有声

Expansion words and phrases（拓展词汇）

white sauce 白少司
remoulade sauce 尼莫利少司
butter sauce 黄油少司
saffron sauce 藏红花少司
Robert sauce 罗伯特少司
French dressing 法国汁

borscht 罗宋汤
goulash soup 匈牙利牛肉汤
onion soup 洋葱汤
consommé de royal 皇家清汤
bouillabaisse 海鲜汤
pureed spinach soup 菠菜蓉汤

Notes（重点词汇、短语）

Task 1

solid food	固体食物
taste better	味道更好
thickened	黏稠的
keep/stop whisking	开始/停止搅拌
ready to serve	准备食用
frothy	多泡的

Task 2

put... into	把……放进
add the oil slowly	慢慢地加油
tablespoon(tbsp)	汤匙
jug	罐、壶
remove the foam	撇去泡沫
to your desired taste	根据自己的口味

Task 3

blender	搅拌器
garnish	点缀、装饰
melt the butter	加热黄油
serve the broccoli with cheese sauce	把芝士少司浇在西蓝花上

Culture life（文化生活）

少司的基本知识

【少司的概念】

少司是英文 sauce 的译音，有的地方译成沙司，是指经厨师专门制作的菜点调味汁。少司的种类和做法很多，有咸、酸、甜、辣等口味。西餐菜肴中，加入不同的少司能使菜式有多种变化。

【少司的种类】

西餐少司的种类很多，在一般情况下可以根据制作菜品的用途来进行分类，主要分为冷菜少司、热菜少司和甜品少司。其中冷菜少司主要用于调制西餐冷菜，热菜少司用来调制西餐热菜，甜品少司用来调制西点。

【少司的作用】

各种少司都是由不同的基础汤汁制作而成的，这些汤汁都含有丰富的鲜味物质，同时还能把各种调味品溶于少司中，使菜肴富有口味，而且大部分少司有一定热度，能均匀地裹在菜肴的表层，这样就能使一些加热时间短，未能充分入味的原料同样富有滋味。

由于在制作一些少司时使用了油脂，因此少司色泽会显得鲜艳光亮，而且在装盘时少司浇淋所形成的图案能够平衡它与主菜的重心。从而彰显主料的特点，增加整体造型的流动感，使菜品的造型更加完美。

【少司菜肴】

Hollandaise sauce atop a salmon eggs Benedict

Samosas accompanied by four sauces

Fish with lemon sauce

Practice 巩固练习

Exercise 1 Translate these words into Chinese or English.（中英单词互译）

1. sauce _____   2. 基础汤 _____

3. curry _____   4. 蓉汤 _____
5. vinegar _____   6. 清汤 _____
7. mayonnaise _____   8. 南瓜汤 _____
9. simmer _____   10. 奶油蘑菇汤 _____

Exercise 2 Tick the different words.（找不同）

1. A. brown sauce   B. cream sauce   C. tomato sauce   D. mayonnaise
2. A. puree   B. stock   C. consommé   D. vinegar
3. A. salt   B. pepper   C. oil   D. sugar
4. A. whisk   B. bowl   C. knife   D. parsley
5. A. sweet   B. spicy   C. sour   D. cream

Exercise 3 Match the key phrases with these pictures.（连线）

Garnish the soup.

Dice the potatoes with a knife.

Boil the potatoes in a pot.

Mix the potatoes with cream chicken stock.

Exercise 4 Write the sentences in the right order.（排序写句子）

1. add, the, you, salt, to, taste, may, your, desired

_____

2. kinds, how many, altogether, of, there, sauces, are

_____

3. garnish, into, pour, soup bowl, and, fresh parsley, with

_____

4. whisking, keep, thickened, is, until, the mixture

_____

5. show, mayonnaise, you, can, me, to, make, how

_____

Exercise 5 Translate western recipe.（翻译西餐菜单）

Russian borscht   Oxtail soup

Stewed fish with tomato sauce

Roasted beef sirloin with red wine sauce

1. 香浓牛尾汤_____

2. 罗宋汤_____

3. 西冷牛排配红酒少司_____

4. 番茄汁烩鱼片_____

Exercise 6 Find the answers.（找答案）

1. When shall I stop whisking?　　　　A. Sure. Go ahead.

2. Shall I cut the potatoes?　　　　　　B. For about 10 minutes.

3. What should I do with the butter?　　C. He is cooking Hollandaise sauce.

4. What is the chef doing?　　　　　　D. Melt the butter in the pan.

5. How long shall I simmer the soup?　　E. Stop whisking when it is thickened.

Exercise 7 Translate the following sentences into Chinese or English.（中英文句子翻译）

1. Whisk until the sauce is thickened and it is ready to serve.
_____

2. Have you got all the ingredients ready?
_____

3. 我需要用到什么厨具吗?
_____

4. 您能告诉我怎么做奶油蘑菇汤吗?
_____

5. 汤姆正在做蛋黄酱。
_____

Unit 9
参考答案

Unit 10
Pastry & Bakery

You will be able to:

1. get familiar with baking materials and ingredients;
2. know the baking tools and usages;
3. talk about the process of making pastries and cookies.

Unit introduction 单元介绍

本单元我们将学习西餐糕点类的相关内容。西餐糕点通常包括蛋糕、曲奇、面包等。通常会用到低筋面粉、黄油、苏打粉等原料。因此本单元的任务是学习制作过程及所用原料的英文表述。

扫码看课件

Thinking map（思维导图）

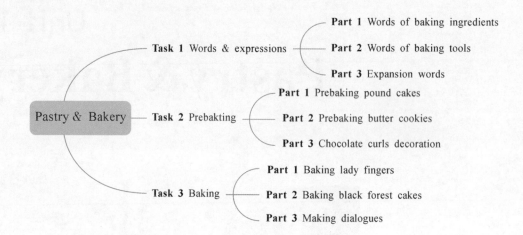

Warming up

Activity 1 Look and match. 看单词，连线。

| cake | pudding | bread | cookie |

外教有声

Activity 2 Listen and tick. 听录音，勾出小神厨要做的西点。

Task 1　Words & expressions 词汇

Part 1　Words of baking ingredients. 烘焙原料单词。

Activity Read and write 读单词,看图写英文。

breadcrumbs 面包屑　　vanilla essence 香草精　　whipping cream 淡奶油　　fish gelatin 鱼胶
cake flour 低筋面粉　　caster sugar 细砂糖　　baking soda 苏打粉　　custard powder 吉士粉

_____

_____

_____

_____

_____

_____

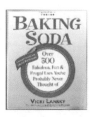

_____

_____

Part 2　Words of baking tools. 烘焙工具单词。

Activity 1 Listen and number. 听录音，标出正确顺序。

（　）scraper 刮刀　　　　　　（　）measuring cup 量杯　　　　（　）spatula 橡皮刮刀
（　）electric mixer 电动搅拌器　（　）scale 秤　　　　　　　　（　）mould 模子
（　）brush 刷子　　　　　　　（　）aluminum foil paper 铝箔纸

Activity 2 Look and translate. 单词互译。

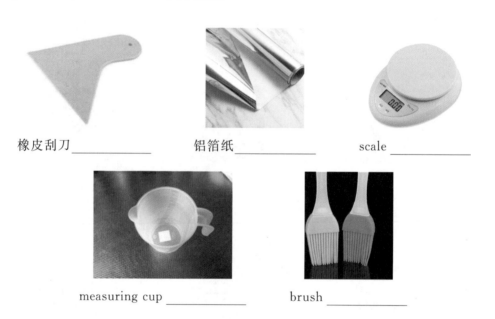

橡皮刮刀_____　　铝箔纸_____　　scale _____

measuring cup _____　　brush _____

Activity 3 Look and write. 根据描述写出单词。

1. _____ It's a hand-held tool that is used for lifting or spreading while baking.
2. _____ It's a kind of material that is used for taking shapes.
3. _____ It is used for mixing dough or flour.

Unit 10   Pastry & Bakery

Part 3   Expansion words. 拓展词汇。

Activity 1 Look and match. 看图连线。

面团　　面糊　　裱花袋　　秤　　裱花嘴　　巧克力碎　　揉　　面粉筛
knead　dough　chocolate chips　pastry bag　pastry syringe　scale　batter　flour sieve

 Activity 2 Listen and write. 听录音写单词。

_____

_____

外教有声

Activity 3 Fill in the missing letters and make a self-evaluation. 补全单词并做自我评价。

1. kn __ d   2. b __ tt __ __   3. p __ try bag   4. s __ le   5. si __ e
6. d __ gh   7. mo __ __ __   8. s __ rin __ e   9. fl __ r   10. cr __ m

Assessment: If you can write 8-10 words, you are perfect.

If you can write 4-7 words, you are good.

If you can write 1-3 words, you will try again.

Task 2 Prebaking 预烘

Part 1 Prebaking pound cakes. 预烘磅蛋糕。

Activity 1 Write down the tools for prebaking pound cakes. 写出预烘磅蛋糕所用工具名称。

_____   _____

Activity 2 Look and choose. 看图并选择。

A. Breaking eggs   B. Measuring the dry cake flour

C. Beating butter, flour and sugar   D. Adding vanilla powder

E. Stirring all the things   F. Pouring the batter

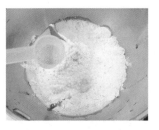  

1._____   2._____   3._____

4._____   5._____   6._____

Unit 10　Pastry & Bakery

Activity 3 Listen and fill in the blanks. 听录音，完成对话。

A. pour the batter into the cake pan　　　B. preheat the oven to 325 ℉

C. flour, sugar and butter　　　D. blend butter and sugar

Chef: Today I'll teach you how to prepare for baking a pound cake.

Jack: Wonderful. What ingredients do we need, chef?

Chef: They are _____.

Jack: How shall we do with them?

Chef: You can _____ with a mixer to make batter.

Jack: I've done it. How about next?

Chef: Add eggs one by one and turn down the speed to add flour slowly.

Jack: Oh, what is the next?

Chef: _____ and get ready for baking.

Jack: I see. I will _____.

Part 2　　Prebaking butter cookies. 预烘黄油曲奇。

Activity 1 Hare a look at the processing. 看烘焙黄油曲奇的准备过程。

Activity 2 Listen and complete the dialogue. 听录音，参考 Activity 1 完成对话。

A. add one a time　　　B. stir butter and salt

C. add sugar　　　D. self-rising flour

Jack: What shall we do with the ingredients for baking, chef?

Chef: We should _____ together with a spoon.

Jack: OK. It's so easy. And then?

Chef: _____ to the mixture and continue stirring until the mixture becomes smooth.

Jack: Shall we add eggs to the mixture next?

Chef: Yes, pay attention to _____ to make sure each one is blended.

Jack: Then, what's the next?

Chef: Mix in _____ with your spoon.

Part 3 Chocolate curls decoration. 巧克力碎卷装饰。

Activity 1 Look and translate. 看图，翻译英文。

1. Melt chocolate over very low heat.
_____

2. Spread melted chocolate into the thinnest layer possible.
_____

3. Scrape chocolate with metal scraper to form chocolate curls.
_____

Activity 2 Make dialogues. 根据所给信息进行对话练习。

A: What should we do next step?
B: We should _____. (combine)

A: What are we doing now?
B: We are _____. (add... to)

A: What shall we _____? (sift, sugar)
B: Yes, we shall.

A: What are we doing now?
B: We are _____. (pour, to the layer)

Task 3 Baking 烘焙

Part 1 Baking lady fingers. 手指饼的制作。

> 手指饼是一款备受青睐的意大利甜食,除了直接食用以外,还可以泡牛奶和蘸巧克力酱吃。它更大的作用是作为芝士、慕斯蛋糕的垫底、夹心和围边。

Activity 1 Read the recipe. 读菜单。

❶ **Ingredients**

egg yolk
vanilla essence
cornstarch
sugar
cake flour
pistachio nuts

❷ **Methods**

1. Beat egg yolk and sugar on high speed until the mixture becomes thick and pale yellow.
2. Add the cream of tarter in the foamy egg white.
3. Fold the white into egg yolk and flour mixture, mixing only until incorporated.
4. Sift together the soft flour and cornstarch.
5. Put the batter into pastry bag.

Activity 2 Tick the ingredients. 勾出所需原料。

☐ onion   ☐ butter   ☐ cake flour   ☐ vanilla
☐ salt   ☐ sugar   ☐ chili powder   ☐ gelatin

Activity 3 Look and write. 看图写出制作流程。

A B C

A. _____
B. _____
C. _____

Part 2 Baking black forest cakes. 烘焙黑森林蛋糕。

> 黑森林蛋糕是德国的著名甜点，德文原意是"黑森林樱桃奶油蛋糕"。它融合了樱桃的酸、奶油的甜和樱桃酒的醇香。巧克力相对较少，以突出樱桃的醇香。

外教有声

Activity 1 Read the dialogue. 读对话。

Jack: We are going to make black forest cakes. What should we do?

Chef: You should prepare the utensils first. What do you think they are?

Jack: Spatula, measuring spoon, electric mixer are necessary.

Chef: Wonderful. Don't forget the tinfoil.

Jack: Oh, I almost forget it. How about the ingredients?

Chef: Eggs, butter, whipping cream, baking flour, cherry and chocolate chips.

Jack: We've finished them, chef.

Chef: Now let's make the dish.

Activity 2 Tool preparation. 工具准备。

橡皮刮刀_____   刀_____   烤箱_____

Unit 10　Pastry & Bakery

量勺_____　　锡纸_____　　打蛋器_____

Activity 3 Ingredient preparation. 原料准备。

鸡蛋_____　　面粉_____　　黄油_____

糖_____　　香草精_____　　巧克力粉_____

淡奶油_____　　樱桃_____　　巧克力碎_____

Activity 4 Look and write the production process. 看图写出制作流程。

1. Carefully measuring out of all the ingredients.
2. Preparing the pan and heating the oven.
3. Melting together the butter and vanilla essence and set them aside.
4. With an electric mixer, combining sugar and eggs.
5. Baking the mixture in the oven for 20 or 25 minutes until it rises.
6. Making the poached cherries.

西餐烹饪英语

_____ _____ _____

_____ _____ _____

Part 3 Making dialogues. 对话练习。

A: What is your favorite cake?
B: It is _____.
A: How do we call it in Chinese?
B: It is called _____.

mousse cake
慕斯蛋糕

chiffon cake
戚风蛋糕

pound cake
磅蛋糕

Necessary words and phrases（必需词汇）

pudding [ˈpudɪŋ] 布丁
mousse [muːs] 慕斯
sponge cake [spʌndʒ keɪk] 海绵蛋糕
muffin [ˈmʌfɪn] 玛芬蛋糕
vanilla [vəˈnɪlə] 香草
breadcrumbs [ˈbredkrʌmz] 面包屑

scale [skeɪl] 秤
scraper [ˈskreɪpə(r)] 刮刀
dough [dəʊ] 面团
batter [ˈbætə(r)] 面糊
crunchy [ˈkrʌntʃi] 易碎的
knead [niːd] 揉、捏

Expansion words and phrases（拓展词汇）

brown sugar [braun ˈʃugə(r)] 黄糖
high flour [haɪ ˈflaʊə(r)] 高筋面粉
pastry bag [ˈpeɪstrɪ bæg] 裱花袋
baking soda [ˈbeɪkɪŋ ˈsəʊdə] 小苏打
blend [blend] 搅拌

soft flour [sɒft ˈflaʊə(r)] 低筋面粉
pastry syringe [ˈpeɪstrɪ sɪˈrɪndʒ] 裱花嘴
custard powder [ˈkʌstəd ˈpaʊdə(r)] 吉士粉
whipping cream [ˈwɪpɪŋ kri:m] 淡奶油

外教有声

Notes（重点词汇、短语）

Task 1

preheat the oven	烤箱预热
measure ingredient	测量所需原料
melt butter	熔化黄油
mix sugar and flour	将糖和面粉搅拌
get rid of	拿出，放弃
forget to do	忘记做……

Task 2

fill with cream	添加奶油
cut down/cut off/cut open	减少/切断/切开

Task 3

combine... and...	把……和……混合
set something aside	把……放在一旁
fluffy egg white	打发蛋白
mix... and... in	把……和……揉成……
sift the flour	筛面

Culture life（文化生活）

蛋糕的传说

【拿破仑——百万层酥皮】

拿破仑蛋糕其实和拿破仑没有关系，说法之一是由于它的英文名 Napoleon，其实是 Napolitain 的误传，指一种来自意大利 Naples 的酥皮名字，至今被写作 Napoleon 而已。拿破仑蛋糕的法文名 Mille-feuille，即为有一百万层酥皮的意思，所以它又被称为千层酥。

【黑森林蛋糕】

"Black forest cake"——黑森林蛋糕,又称"黑森林樱桃奶油蛋糕",是一种鲜奶蛋糕。相传很久以前,德国南部有一处名为黑森林的旅游胜地,盛产黑樱桃。每年樱桃丰收之季,当地的人把过剩的黑樱桃夹在巧克力蛋糕内,或是将其一颗颗细心装饰在蛋糕上。在制作蛋糕的鲜奶油时,也会加入不少的樱桃汁。这种蛋糕从黑森林地区传出去后,便成了黑森林蛋糕。

Practice 巩固练习

Exercise 1 Translate these words into Chinese or English.(中英单词互译)

1. strong flour _____
2. vanilla essence _____
3. baking soda _____
4. sweet chocolate _____
5. sponge cake _____
6. 奶油奶酪 _____
7. muffin _____
8. 细砂糖 _____
9. scraper _____
10. 揉 _____

Exercise 2 Tick the different words.(找不同)

1. A. strong flour   B. soft flour   C. custard powder   D. butter
2. A. egg yolk   B. egg white   C. sugar   D. tin paper
3. A. mixer   B. spatula   C. scraper   D. flour
4. A. stir   B. heat   C. melt   D. spatula
5. A. cream   B. vanilla   C. butter   D. cherries

Exercise 3 Match the key phrases with these pictures.(连线)

mix ingredients   bake cake   mould cake   decorate cake

Exercise 4 Write the sentences in the right order.(排序写句子)

1. the almond cookies, to, 1 hour, until, 45 minutes, they, please, brown golden, bake, are, from

2. shall,mix,I,speed,the,high,ingredients,at

_____

3. the,on,the,of,place,the,bottom,pan,dough

_____

4. the,mixture,divide,into,balls,walnut,size

_____

Exercise 5 Translate western pastry.（翻译西点菜单）

Almond slices   Swiss fruit roll

Onion bacon quiche   Baguette

Dark chocolate mousse

1. 洋葱培根塔_____   2. 杏仁片曲奇_____

3. 瑞士水果蛋卷_____   4. 法棍面包_____

5. 黑巧克力慕斯_____

Exercise 6 Find the answers.（找答案）

1. Place a layer of it on the pan.   A. scraper
2. Preheat it to 180 °F.   B. parchment paper
3. Mix eggs with sugar and flour.   C. mixer
4. Cut and shape the dough.   D. oven
                                            E. measuring cup

Exercise 7 Translate the following sentences into Chinese or English.（中英文句子翻译）

1. Cool the almond cookies on a rack before serving.

_____

2. I will pour the ingredients into the baking pans.

_____

3. 您需要我烤面包吗？

_____

4. 面点师傅教我如何对角地切奶酪蛋糕。

_____

Unit 10
参考答案